Chavan vs Einstein: The Universe Was Never Curved

Sandeep Chavan

Published by SANDEEP CHAVAN, 2025.

CHAVAN VS EINSTEIN: THE UNIVERSE WAS NEVER CURVED

First edition. June 8, 2025.

ISBN: 979-8231156856

Written by Sandeep Chavan.

Table of Contents

To Albert Einstein—

Not as a rival,but as a ripple before me.

You dared to bend realitywhen the world still clung to straight lines.You questioned clocks, motion, and massuntil space itself began to speak.

This book is not against you.It is because of you.

Because you asked the right questions—even when the answers weren't ready.

Because you stood at the edge of the knownand didn't flinch when it dissolved.

Because you reminded usthat imagination is more important than knowledge—and alignment, perhaps, more important than either.

May this work honor your spiritnot by preserving your equations,but by continuing your courage.

Where your light curved,this ripple realigns.And still—we walk together.

"We cannot solve our problems with the same thinking we used when we created them."— *Albert Einstein*

"We didn't need better answers. We needed a better field to think in."— *Sandeep Chavan*

Disclaimer

This book is not affiliated with, endorsed by, or reviewed by any official scientific institution or academic authority. It is an independent work of research, interpretation, and philosophical inquiry, authored with the intent to challenge, rethink, and realign the foundations of modern physics through a new lens: **Universal Energy Dynamics (UED).**

While references to Einstein's theories are made throughout, this book does not intend to discredit his contributions, but rather to extend the conversation he began—and to do so with full respect for the legacy he left behind.

The contents presented here—**ripple logic, SVE, ATT, and field-based reality**—are alternative frameworks, rooted in the author's personal research, experience, and philosophical reasoning. These ideas are not yet part of conventional scientific consensus and should be approached with open curiosity, not blind acceptance.

This book is also not a substitute for formal education or professional scientific guidance. It is meant to inspire thought, dialogue, and deeper perception.

Use it not as doctrine—but as invitation. Not to believe—but to perceive.

Opening Note

A message from Sandeep Chavan — why this book had to be written, and why now.

"I did not set out to challenge Einstein. I set out to understand the universe. The challenge was just a consequence."

This book wasn't born from rebellion. It was born from a growing silence—a quiet, persistent discomfort with the way we've come to define reality. For years, like many in the scientific community, I stood in awe of Einstein's intellect. His thought experiments were more than clever—they were revolutionary. But as I dug deeper, questions began to ripple through the certainties.

Why does time bend, but never break? Why must light obey a universal speed limit, even across empty space? Why do we build entire theories on coordinates, dimensions, and assumptions—without asking what *space* really is?

And most haunting of all: Why do we call it a "Theory of Everything" when it explains so little?

I am not a prophet, a physicist by title, or a rebel in a lab coat. I'm a teacher. A seeker. A man who, like you, lives in this universe and wants to understand it—not through borrowed equations, but through logic that feels real.

The framework you're about to encounter—**Universal Energy Dynamics (UED)**—is not a rejection of Einstein. It's an invitation to evolve. It doesn't dismantle his genius. It completes it.

Where Einstein saw a curved spacetime, I see ripple fields in a structured vacuum. Where GR describes effects through geometry, UED traces causes through consequence. Where physics split into relativity and quantum mystery, UED brings them back—into a unified ripple reality.

So why now?

Because the world is waking up. Old dogmas no longer hold. The ivory towers are cracking, and curiosity is becoming contagious again. Young minds are not just asking "how," they're asking "why."

This book is for them. For thinkers who feel something's missing in mainstream science. For dreamers who want reality to make sense again.

We live in a time when information floods us, but understanding escapes us. This book is not just about equations—it's about truth, intelligence, and the courage to see differently.

The universe was never curved. It ripples. And it's time we learn to listen.

— Sandeep Chavan

Independent Physicist India, Developer of UED Framework 2025

Preface

Why This Book Had to Be Written

There are moments in life when a question refuses to leave you. Not because it's difficult— but because something inside you already knows the answer, and is just waiting for the world to catch up.

This book began with such a question: What if we misunderstood the very foundation of reality—not because we lacked intelligence, but because we lacked alignment?

For over a century, physics has worshipped equations, forces, and frames. It gave us astonishing tools—rockets, GPS, black hole images. And yet, it left untouched the deeper ripples of existence:

- Why does perception shape time?
- Why does gravity feel effortless, yet remain mysterious?
- Why do we keep inventing dark things to patch bright ideas?

In the silence beneath those questions, a different voice emerged. Not the voice of opposition—but of completion. Not a new theory, but a new lens.

This is not a book against Einstein. This is a book with him, beyond him, beside the very questions he couldn't finish asking. It's not about dismantling Relativity—it's about re-centering Reality.

Through the lens of Universal Energy Dynamics (UED), we return to the beginning—not of the universe, but of our understanding. We replace the myth of spacetime with the rhythm of ripples. We trade motion for alignment. We shift from force... to consequence.

This book was written for:

- The physicist who feels something's missing, even when the math checks out.
- The philosopher who wants truth without mysticism.
- The student tired of memorizing frameworks that don't feel alive.
- The silent observer who always sensed there was more.

And most of all— for the next generation who will not inherit models, but reality itself.
May this work invite not belief, but resonance. May it not impress you, but align you.
Not because it's perfect— but because it dares to ripple.
— *Sandeep Chavan*

How to Read This Book

A Field Guide for the Aligned Mind

This is not a textbook. There are no problem sets at the end. No multiple-choice answers. No requirement for prior knowledge in physics.

Because this isn't a book that teaches. It's a book that reminds.

Read With Openness

You may not agree with everything. You're not supposed to. This book doesn't ask for belief—it asks for resonance. When something rings true, pause. That moment is more important than any page number.

Go Slow

This isn't a race. Some chapters will feel like science. Some will feel like philosophy. Some may feel like poetry. And some may simply feel like you remembering what you forgot.

Read them as you would observe ripples— gently, patiently, with attention.

Don't Force Understanding

If something doesn't make sense right away, that's not confusion—it's realignment in progress. Let the ideas settle. Your mind isn't a machine. It's a field. Clarity will come—not from memorizing, but from sensing.

Use the Margins for Silence

Highlighting is fine. But sometimes, leave space blank—so your own ripple can emerge.

This book doesn't just explain things. It listens to you as you read. And sometimes, the most important thing you'll find is the thought you didn't know you had—until now.

Revisit, Don't Review

This is a book meant to be re-entered, not just remembered.

Every rereading reveals a different layer— not because the book changed, but because you did.

Final Suggestion:

Don't just read it. Sit with it. Let it ripple.

If it changes your thoughts, beautiful. If it only changes your pause, that's even more powerful.

Part I: The Curved Illusion

Chapter 1: The Idol of Curvature

How 20th-century physics worshipped geometry and lost its grounding
"At some point in the 20th century, we stopped asking what the universe is made of—because we were too busy admiring the shapes it made."

It began as a bold insight: gravity, the invisible force that tethered planets and bent the paths of stars, wasn't a force at all. It was the shape of space. A dent. A curve. A warping of the stage itself. And with that single metaphor, physics crowned geometry as its god.

No longer did we speak of fields, flows, or substance. We spoke of fabric, metrics, tensors—of curvatures elegant enough to be worshipped, but distant enough to never be touched. Equations replaced essence. Coordinates became more real than the canvas they described. And in our awe of mathematical beauty, we forgot to ask one simple question:

What exactly is bending?

We applauded General Relativity not just because it worked—but because it was beautiful. But beauty, when left unexamined, becomes dogma. And dogma, in science, becomes a prison.

This chapter is not a condemnation of Einstein. It is a confrontation with the culture that followed him—a culture that mistook symbolic elegance for fundamental truth. It is the beginning of a deeper conversation. One where we stop bending space, and start listening to what ripples beneath it.

Geometry as God

For thousands of years, geometry was not just a tool of mathematics—it was a philosophy of order. The Greeks believed in a cosmos shaped by perfection. Euclid's geometry didn't just describe lines and angles—it described a world they believed was built from divine proportions. Geometry gave us circles, symmetry, and the illusion that truth could be drawn with a compass.

This reverence never really faded. When Newton came along, his mechanics were grounded in space and time as absolute stages—a geometric stage, on which motion was choreographed with force. But Newton still believed in substance—that space and time were real, but passive—backgrounds against which energy moved.

Then came Einstein, and with him, a transformation more philosophical than physical. His insight was stunning: mass doesn't just exist in space—it shapes it. Gravity, long thought to be a force pulling from a distance, was suddenly replaced with curved geometry. Planets didn't fall—they followed geodesics. The Sun didn't pull—it dented the cosmic stage.

"Space tells matter how to move; matter tells space how to curve." – John Archibald Wheeler, summarizing Einstein

This poetic line became scripture. Physics was no longer about what existed—it was about how it bent.

Suddenly, geometry wasn't a language—it was the law.

Einstein didn't just give us a theory. He gave us a new idol—one that looked nothing like the material world we could touch. Instead, it was a stage made of equations, infinitely smooth, elegant, continuous, and invisible. And we fell in love with it.

Elegance Mistaken for Essence

The equations of General Relativity were not just accurate—they were aesthetically mesmerizing. Riemannian geometry, tensors, curvature invariants—they painted the universe as a symphony of form. But in that elegance, something quietly slipped away: ontology.

No one could say what spacetime *was*. It had no substance, no field, no interaction—only shape. We assumed it could bend, ripple, even collapse into singularities, and yet we had no idea what was bending. We accepted curvature as cause, even though it looked suspiciously like an effect.

"The theory is beautiful," they said. "But what is space?" someone asked. "It's what curves," they answered. "What curves?" "Spacetime." "But what *is* spacetime?" Silence.

Physicists like Dirac, Wheeler, and Penrose spoke often of the *beauty* of the mathematics, the *purity* of the form, the *inevitability* of the equations. But almost none spoke of substance. None asked: *Is there a medium? A field? A structure beneath these forms?*

Because in this new religion of curvature, questions of substance became heresy.

Geometry Replaces the Ground

By the mid-20th century, geometry had fully replaced grounding in physics. Everything—gravity, motion, even time—was reduced to coordinates, frames, and manifolds. But geometry, by nature, is descriptive. It tells us how things relate—not why they exist. It maps the terrain, but it doesn't explain the earth beneath.

And yet, for over a century, we built on this hollow stage, treating description as causation. We believed curvature explained motion. That equations explained force. That elegance implied truth.

In doing so, we lost sight of the deeper question: What causes the shape in the first place?

Core Insight: A Forgotten Foundation

When we crowned geometry as the god of physics, we silenced an older, more intuitive question:

Is there something beneath the shape?

UED says yes.

It brings us back to ontology. It reminds us that shape arises from substance. That motion arises from ripple. That what we observe—bending, distortion, time delay—is not geometry in action, but energy in consequence.

Geometry is not the cause of gravity. It's the footprint of something deeper. A ripple in the field—not a wrinkle in a fabric.

And that is where we begin our departure. Not by· dismissing Einstein's brilliance—but by recognizing the assumptions he left unquestioned. The next chapter takes us directly into those assumptions, and shows how even his most elegant thought experiments never escaped the confines of curvature—but hinted at something far more real:

The ripple underneath.

From Gravity to Geodesics

When physics replaced interaction with curvature—and called it truth.

Before Einstein, gravity was strange—but at least it was intuitive. Newton saw it as a universal force: invisible, constant, and instantaneous. Every object with mass pulled on every other object, no matter the distance. It was mysterious, yes—but it worked. It predicted the motion of planets, explained tides, and shaped the mechanical worldview of the universe.

But Newton's gravity had one problem: it assumed action at a distance without a medium. He admitted as much. He refused to guess what carried that force across empty space. "I frame no hypotheses," he famously said.

This uncertainty lingered for centuries—until Einstein stepped in, and did what no one expected:

He abolished gravity.

Einstein's Leap: From Pull to Path

Einstein didn't fix Newton's gravity. He deleted it and replaced it with something conceptually radical. Objects no longer pulled on each other. Instead, mass told space how to curve, and space told objects how to move.

Gravity was no longer a force—it was the illusion of a straight path in curved spacetime. If that sounds strange, it's because it was.

Bodies now followed geodesics: the "straightest" lines in a curved space. Free fall wasn't falling anymore—it was the most natural path available, just distorted by the shape of the background.

A revolution was born. And so was a profound assumption that would shape physics for a century:

That spacetime curvature is real. That curvature explains motion. That geometry is the cause, not just the map.

The Equation That Redefined Everything (But Explained Nothing)

Einstein's field equations were beautiful. They balanced energy and momentum with curvature, packaged in a framework of tensor calculus. They were hailed as the most elegant equations in science.

But in all their beauty, they never said what spacetime is. Only that it bends.

This is where physics took a dangerous detour. The elegance of the equation seduced the scientific world. We didn't just accept the math—we accepted the geometry as physical reality, without questioning what it was made of.

We celebrated a theory that bent reality—but never asked, *what is it bending into?* What is the medium? What is the substrate? What ripples when spacetime curves?

The Rise of Curvature Worship

The success of GR was undeniable. It predicted phenomena that Newtonian physics couldn't: the precession of Mercury's orbit, the bending of light during a solar eclipse, gravitational time dilation. Each confirmation made Einstein more than a scientist. He became a symbol of ultimate truth.

Physicists weren't just impressed—they were entranced.

"One of the greatest achievements of human thought." – Max Born
"An advance as profound as Copernicus." – Arthur Eddington

But in our praise, we stopped probing. Geometry became sacred. Spacetime became unquestionable. We convinced ourselves that curvature was the answer, and forgot to ask if it was just the appearance.

The Critical Error: Mistaking Description for Explanation

Curvature described what gravity *does*—it didn't explain *why* gravity happens. The math fits. The predictions match. But if you strip away the symbols and ask what's actually occurring beneath the curves—there's nothing. Just a framework.

The fatal assumption was this:

That curved spacetime is a physical thing. That geometry isn't just math—it's mechanism.

That assumption became law. And once physics replaced interactions with paths, and forces with frames, it stopped seeking cause. It settled for consequence.

Did Curvature Explain Gravity — or Just Repackage It?

This is the core question.

Did General Relativity truly explain gravity? Or did it simply redefine its language, wrapping the same mystery in more beautiful math?

The geodesic path may be elegant. The predictions may be precise. But the field equations never told us what space is, what time is, or what bends when we say spacetime curves.

It told us *how things appear*. It never told us *what makes them so*.

That's the fracture UED addresses. It reclaims the medium—Structured Vacuum Energy—and shows that what we mistook as curvature is, in fact, a ripple consequence. Not form, but function. Not bending, but alignment.

In the next section, we'll explore how this worship of curvature became a cultural phenomenon in science—and why physics stopped asking the deeper, more dangerous questions.

The Consequences of Worship

How modern physics lost its center chasing curved abstractions.

When Einstein curved the universe, the world of physics celebrated. But that celebration soon turned into silence—the kind that fills a room when too many people agree without knowing why.

Because something peculiar happened in the aftermath of General Relativity's success: physics lost its foundation. It had equations, predictions, and a mathematical scaffolding that could launch satellites and adjust GPS clocks. But it lacked something far more crucial—a unified picture of reality.

And the deeper cost? We didn't just admire curvature. We built a temple around it.

A Universe That Wouldn't Curve

While General Relativity reshaped gravity as curved geometry, quantum mechanics refused to play along. At the smallest scales of nature—atoms, particles, and fields—geometry meant nothing.

Quantum theory operated on probability, uncertainty, and discreteness. It spoke a different language entirely:

- Particles were excitations in fields, not masses bending space.
- Forces emerged from quantum interactions, not from geodesics.
- Time was a parameter—not a flexible, stretchable fabric.

Attempts to bend the two frameworks into a single whole were frustrating at best, futile at worst.

You could curve a spacetime. You could quantize a field. But you couldn't quantize a curve—and you couldn't curve a quantum field.

It was as if the universe had been split into two books, written in two different alphabets, by two authors who had never met.

This fracture between GR and quantum mechanics wasn't just a technical problem. It was an ontological incompatibility. They didn't disagree about results—they disagreed about what reality is.

The Rise of Patchwork Physics

Instead of stepping back and asking whether the assumptions were wrong, physics piled on more complexity. To make GR work with quantum theory, we invented entire theoretical scaffolds:

- **String Theory:** claimed particles were tiny vibrating strings in 10 or 11 dimensions—unseeable, untouchable, and forever unverifiable.
- **Dark Matter & Dark Energy:** inserted as invisible patches to fix gravitational behavior that GR couldn't explain.
- **Quantum Gravity:** an ever-shifting collection of mathematical bandages trying to stitch together two irreconcilable logics.

None of these were grounded in a known physical medium. They were ideas *about* ideas, floating in conceptual space. Their elegance was admired—but their explanatory power remained speculative.

We left behind reality to chase mathematical beauty. We replaced substance with symbolism.

And the result? A century of theory with no closure.

Equations vs. Understanding

It must be said clearly: Experimental success does not imply ontological truth.

Just because General Relativity predicts the bending of starlight doesn't mean spacetime is literally a curved fabric. Just because your GPS needs Einstein's corrections doesn't mean reality is four-dimensional geometry. These are phenomenological matches, not proof of what the universe *is*.

A thermometer tells you the temperature. It doesn't tell you what heat *is*. General Relativity tells you what objects will do. It doesn't tell you *why* gravity happens.

This is where physics made its deepest mistake. It mistook predictive accuracy for existential explanation. It turned coordinate transformations into physical laws. It worshipped curvature not because it was right—but because it *worked well enough.*

The Medium We Forgot

All the while, the real question sat in the corner, ignored:

What is space? What is time? What is the field that allows anything to exist, move, or interact?

No answer came. Not from geometry. Not from strings. Not from quantum foam.

The tragedy is not that we were wrong—but that we stopped looking for what might be *right.* We were so mesmerized by curved equations that we forgot to ask if there was a substrate—a real medium—beneath it all.

That medium is Structured Vacuum Energy (SVE), and its behavior is not curved—it ripples. Motion, force, and time are not drawn from geometry—they are consequences of ripple interactions. The field is not empty. It is alive with tension, balance, and feedback.

The Cost of Elegance

For over a hundred years, physicists have refined equations, built colliders, measured starlight, and stretched the mathematics of GR to its limit. And yet:

- We still don't know what gravity *is.*
- We still can't unify the big and the small.
- We still can't define time in a way that makes sense across both frameworks.

- We still can't explain why space "bends"—only that it appears to.

We've mastered the symbols but forgotten the song.

And now, it's time to listen again—not to the curves we drew, but to the ripples we ignored.

In the next section, we'll trace how this silence gave rise to misplaced certainty, and why it's not too late to return to reality—not as curved spacetime, but as structured energy in dynamic harmony.

The Return of Substance

Restoring the forgotten foundation of physics.

When physics began, it wasn't shy about substance. The early thinkers—Democritus, Newton, Faraday, Maxwell—believed in something. Whether it was particles, ether, or electromagnetic fields, they assumed that behind every force was a medium. Something existed. Something moved. Something caused the effects we observed.

But somewhere between Maxwell's waves and Einstein's curves, we lost the medium.

We became content describing the behavior of space, without asking what space was. We measured the effects of time dilation, without questioning what time was made of. We predicted motion, force, and light, without identifying any substrate that carried them.

Physics had shifted from substance to abstraction. And that shift came at a cost.

Bringing Back the Ground Beneath Our Feet

It's time to reintroduce substance to the conversation. Not as a regression to old ideas, but as a return to first principles—the idea that the universe must be made of something real, structured, and dynamic.

That "something" is what UED calls Structured Vacuum Energy (SVE).

SVE is not a void. It's not nothingness. It is the background field of reality, stable yet responsive, silent yet consequential. It is not curved—it is structured. And when disturbed, it doesn't warp—it ripples.

These ripples, not geometric distortions, are the true foundation of what we call motion, time, light, and gravity.

- Gravity isn't a dent in space—it's the result of ripple gradients in a structured field.
- Light isn't a particle traveling—it's a ripple alignment event reaching the observer.
- Time isn't a dimension—it's an emergent delay, a consequence of ripple interaction thresholds.

Geometry is a Shadow, Not a Cause

General Relativity treats geometry as fundamental. It claims that the shape of spacetime *causes* what we observe. But what if that shape is simply a side effect—the shadow cast by deeper dynamics?

UED proposes a reversal: Geometry is not the engine—it's the dashboard. It doesn't drive the universe—it displays its outcomes. The real engine is ripple behavior in SVE.

In this light, the predictions of GR—light bending, time dilation, geodesics—are not wrong. They're reframed. They are ripple consequences, not curvature causes.

We didn't need to reject Einstein. We needed to reconnect him—to a substrate he never named, but which his thought experiments were silently gesturing toward.

A Thought Experiment Revisited

Einstein built his theory from thought experiments, not equations. He imagined trains, clocks, falling elevators, and beams of light. These mental models gave birth to relativity.

But what if those experiments contain clues he didn't recognize? What if the bending of light was never about curved space—but ripple distortion in a dense field? What if simultaneity was not a function of velocity—but of alignment latency between ripple observers?

In the chapters ahead, we will revisit each of these iconic experiments—not to dismiss them, but to decode them anew, through the lens of ripple dynamics and SVE.

Closing Insight

Physics didn't go astray because of Einstein's mistakes. It drifted because of the world's worship of his metaphors—and the failure to ask what lay beneath them.

Einstein offered a map of elegance. UED offers a return to the terrain.

"Einstein's genius wasn't wrong—it was just incomplete. The idol wasn't the man. It was the assumption that reality could be curved without asking what was curving."

Let's now step into those thought experiments—one by one. And this time, let's bring the field with us.

Chapter 2: Einstein's Mind Theater

Train rides, falling elevators, and solar eclipses—where relativity was born.

"You don't need a telescope to change the universe—just a chair, a quiet room, and the courage to ask, 'What if?'"

Before satellites orbited and clocks ticked at relativistic speeds, before black holes became Instagram icons of science, there was a man imagining lightning bolts, train cars, and elevators in free fall. With nothing more than his mind, Albert Einstein reshaped reality. He didn't rely on machines—he relied on paradox. He didn't demand proof—he demanded coherence. And in his theater of thought, where motion, light, and time were the actors, modern physics found its new script.

These thought experiments—simple, visual, intuitive—became the cornerstones of both Special and General Relativity. They painted time as fluid, space as elastic, and gravity as shape. They inspired generations. They were elegant, persuasive, and unforgettable.

But they were also built on assumptions no one dared to question.

Assumptions about light. Assumptions about time. Assumptions about space as a real thing that can bend—but never break.

This chapter isn't about dismantling Einstein's genius. It's about stepping into his thought experiments, not as admirers, but as examiners. It's about peeling back the brilliance to see what truths they revealed—and what truths they buried.

Because if Einstein's thought experiments gave birth to relativity, then maybe, just maybe, a second look can give birth to something more.

Something real. Something rippling. Something deeper than geometry.

Let's return to the train. Let's enter the elevator. Let's watch the stars curve.

But this time, let's ask: *"Was it really space that bent—or was it something else aligning beneath it?"*

The Theater of Thought

Where reality was rewritten—without a lab.

Before the Large Hadron Collider. Before the Hubble Telescope. Before gravitational wave detectors and atomic clocks ticking in orbit—there was a man with a pen, a chair, and a mind unwilling to accept the ordinary.

Albert Einstein changed the course of physics not with experiments—but with imagination.

In an era dominated by instruments and data, he chose daydreams. In place of evidence, he leaned on intuition. And against the backdrop of Newton's towering legacy, he asked the most daring question possible:

What if the world doesn't work the way we've always believed?

While most physicists were building models from measurements, Einstein built reality from contradictions. He distrusted the ease of established knowledge. Where others saw facts, he saw gaps. Where others took motion for granted, he wondered: *what does it mean to be still?*

Imagination Over Apparatus

Einstein once said, *"Imagination is more important than knowledge."* He meant it.

He believed that the greatest leaps in understanding would never come from instruments alone, but from the capacity to think differently about familiar things—to see the hidden assumptions that knowledge forgets to question.

His greatest tools were clarity, curiosity, and contradiction. Using them, he created mental models so vivid, so self-contained, they didn't need test tubes—they needed only a willing mind.

These weren't idle musings. They were the genesis of relativity itself.

The Three Acts of Einstein's Mind Theater

Like a master playwright, Einstein staged three acts that would redefine physics.

Act I: The Lightning Strikes on a Moving Train — The Simultaneity Problem

Two lightning bolts strike the front and back of a train. An observer on the platform sees them as simultaneous. An observer on the train—moving relative to the strikes—does not.

Einstein's conclusion: Simultaneity is not absolute. Time is relative to the observer's motion. This mental scene birthed Special Relativity.

Act II: The Elevator in Free Fall — The Equivalence Principle

Imagine being in a sealed elevator that is either floating in deep space or falling freely under gravity. Could you tell the difference?

Einstein said no. To the person inside, both scenarios feel the same.

This was his Equivalence Principle—the cornerstone of General Relativity. Gravity and acceleration were indistinguishable, so perhaps gravity was not a force at all—but a feature of motion in a curved space.

Act III: The Starlight That Bends — Gravity and Light

If gravity affects everything with energy, shouldn't it also bend light? Einstein predicted that starlight would curve as it passed near a massive object, like the Sun. And in 1919, during a solar eclipse, this prediction was confirmed by Arthur Eddington's famous expedition. This "proof" catapulted Einstein to global fame and sealed General Relativity's authority.

What Made These Experiments Powerful

These weren't random ideas. They were carefully constructed paradoxes—designed to expose the weak points of classical physics and propose an entirely new structure.

But more importantly, they worked because they felt right. They were simple, visual, and compelling. You didn't need a PhD to follow the logic. You just had to sit with the picture long enough to feel the contradiction unfold.

Einstein's genius lay not just in solving problems—but in framing them so clearly that no one could ignore them again.

But What if They Were Built on Unstated Premises?

Here lies the turning point of this book.

While these thought experiments are revered, they are rarely revisited. They are taught as truth, not as constructs. But like all constructs, they rest on assumptions—many of which were never explicitly questioned:

- That light is an entity that travels in straight lines through empty space.
- That motion defines time, rather than time defining motion.
- That space can bend, without asking what *space* actually is.
- That gravity and acceleration are truly indistinguishable—and not just perceptually similar.

These thought experiments were brilliant—but they were not ontologically neutral. They embedded within them a view of reality still chained to classical notions, subtly adapted to fit a new geometry.

And that's the fracture UED exposes. The ripple-based view doesn't deny these experiments—it reinterprets them from the ground up. It restores substance, inserts the medium, and replaces curvature with consequence.

In the chapters that follow, we'll walk through each of these iconic mental scenes. But this time, we won't look for what Einstein saw.

We'll look for what he missed.

The Train and the Lightning – The Myth of Simultaneity

Where time was bent—but perception was never examined.

Let's begin with the most famous ride in the history of theoretical physics—a train moving smoothly along the tracks, and lightning striking both ends.

Einstein's thought experiment was simple but revolutionary. It involved two observers—one on the platform, standing still; the other on the train, moving steadily forward. Two lightning bolts strike the tracks—one at the front of the train, the other at the back—equidistant from the midpoint of the train.

To the observer on the platform, the bolts appear to strike simultaneously. But for the person on the train—who is moving toward one flash and away from the other—they do not appear at the same time.

Einstein's conclusion?

Simultaneity is not absolute.

It depends on the motion of the observer. Therefore, time itself is relative to your frame of reference.

This was the seed from which Special Relativity grew.

If two observers can disagree about whether two events happened at the same time, then time cannot be universal.

It must bend. It must shift. It must adapt to movement.

That's the classical conclusion.

Now let's revisit this scenario—not with admiration, but with a new question:

What exactly is happening in the mind—and field—of the observer?

What Einstein Saw, and What He Assumed

Einstein assumed that light travels—as an independent entity or wave—through empty space, from the lightning point to the observer's eyes. He believed that motion affects the timing of when the light reaches each observer, hence altering their perception of simultaneity.

But what if light is not something moving through space? What if light is a ripple alignment event—an instantaneous resonance between the source field and the observer's field?

What if what appears to be "arrival time" is really perception latency?

UED's Ripple Reinterpretation: The Field Feels First

Under Universal Energy Dynamics (UED), light is not a particle or wave zipping across a vacuum. It is a ripple of alignment within the Structured Vacuum Energy (SVE) field. The observer does not "receive" the light in the classical sense—they register the alignment when their field reaches sufficient coherence with the originating ripple.

Each observer has a unique alignment threshold time (ATT)—the minimum time it takes for their field to register a ripple from an external source. This threshold is not based on motion—it is based on:

- Local ripple tension
- Observer's internal state
- Directional alignment with the source event

- Field interference and distortion around them

In Einstein's train, the observer standing still has a balanced field. The ripple from both lightning bolts reaches alignment threshold at the same time.

The moving observer is not just traveling through space—they are disrupting their field alignment with one end and enhancing it with the other. This causes the ATT for each lightning event to differ—not because of movement, but because of relative field misalignment in the ripple structure.

Simultaneity isn't distorted by motion. It's shaped by the observer's alignment with the ripple event.

Perception, Not Time, Is What Changes

Einstein's model says the time of the events is in dispute because time itself is relative.

UED says: Time never changed. The events occurred simultaneously in the ripple field—but were perceived differently because each observer had a different alignment profile.

This is not a matter of disagreement in clocks. It's a disagreement in when perception becomes possible.

Time, in UED, is not a river that flows faster or slower. It is a delay between energy alignment and field recognition. It's a measure of when the ripple makes contact with the observer's inner field structure—not when the ripple was sent.

Thus, the entire foundation of relativistic time dilation becomes a misinterpretation of delayed perception.

The Fatal Assumption: Light as a Traveler

Everything hinges on this:

Einstein believed light must travel. UED reveals that light doesn't travel—it aligns.

That shift redefines everything.

It means:

- Space is not a passive arena for light to traverse—it's a dynamic ripple field.
- Time is not stretched or compressed by motion—it is emergent from alignment latency.
- Relativity isn't needed to explain the delay—consequence-based ripple perception does the job better.

And crucially—it does so without breaking causality, or invoking multidimensional coordinates.

Ripple Insight

"Time doesn't differ because clocks move. Time differs because perception emerges from field realignment." — UED Principle of Temporal Consequence

This single shift—from traveler to alignment—reveals that Einstein wasn't wrong in what he observed.

He was simply unaware of what was really happening beneath the stage. He thought he was watching the actors.

But UED tells us: He was watching the ripples.

And they were always there—waiting to be felt, not measured.

The Elevator and the Falling Man – The Gravity Illusion

What if gravity is not a force, not a curve, but simply a ripple you fall into?

Einstein was famously skeptical of conventional thinking. He often said the path to deeper truth begins not with complex formulas, but with simple questions. And one of his most powerful questions came in the form of a sealed elevator.

Imagine a person trapped inside an elevator cabin. If the elevator is in deep space, far from any planets or stars, and it begins to accelerate upward at exactly 9.8 m/s^2, the person will feel their feet press against the floor. They'll think they're standing on Earth. Objects will fall, dropped from the hand, just like they would in a gravity field.

Now, imagine the same elevator is free-falling under Earth's gravity, perhaps in a vertical shaft or falling from orbit. Everything inside appears weightless. The person floats. Objects hover or drift. They might think they're in space—even though they're surrounded by gravity.

This mental experiment led Einstein to one of his most important insights:

You cannot distinguish between gravity and acceleration by local observation alone.

From this, he formulated the Equivalence Principle—the idea that gravity and acceleration are, at least locally, the same phenomenon.

That insight became the foundation of General Relativity, leading to the bold proposition that gravity is not a force, but a manifestation of spacetime curvature.

Objects don't fall because they're pulled—they fall because they follow the straightest path in a curved geometry. This was the death of Newton's invisible hand and the birth of geodesics.

But what if that leap, though mathematically profound, was conceptually misplaced?

What if gravity and acceleration aren't equivalent—because neither of them are primary phenomena?

What Einstein Got Right—and What He Missed

Einstein's genius was in relating the observer's experience to a deeper principle. But what if both the experience and the principle rested on an illusion—the illusion of force, the illusion of movement, and the illusion of curvature?

Under Universal Energy Dynamics (UED), gravity is not a mystery. It is not a force. And it certainly is not a geometric side effect.

Instead, UED posits that everything exists within a Structured Vacuum Energy (SVE) field—a dynamic, ripple-responsive medium. Mass does not bend spacetime—it disturbs the SVE, creating ripple gradients. These gradients do not "pull" objects. Instead, they offer low-tension alignment paths that any object, left undisturbed, will passively drift into.

This drift feels like falling, but it is not acceleration. It is field realignment.

Free Fall as a Ripple Drift

When the elevator is in free fall, nothing is "falling" in the classical sense. The cabin and everything inside it are simply surrendering to the ripple gradient. There is no external force. There is no geometric curvature. There is only a structured field and the natural tendency of mass-energy to align with zones of lower ripple resistance.

Likewise, the elevator in space, accelerating at 9.8 m/s^2, is not mimicking gravity. It is generating a ripple resistance, a kind of temporary tension mismatch between the field and its structure. The body responds by pushing against it, creating the illusion of weight.

Einstein saw the effects—but assumed the equivalence implied identity. UED shows that the field causes are fundamentally different.

In free fall, you are relaxing into alignment. In acceleration, you are fighting against misalignment.

The Ripple Landscape Analogy

Imagine the SVE field like an invisible ripple landscape. A mass, like Earth, creates a gentle slope in this field. When you "fall," you are not being pulled—you are sliding.

But here's the twist: you don't feel the slide unless there's resistance. Inside the falling elevator, there is no resistance. Everything aligns with the slope together. Hence: weightlessness.

But stand on Earth's surface, and the ripple gradient is blocked by the floor. Your field wants to move down the ripple slope, but the structure pushes back. That tension? We call it weight.

You never feel gravity itself. You only feel when something stops you from falling.

This is the Ripple Insight: Gravity is not an external force acting on you. It is the ripple field's invitation to realign. And free fall is just your body accepting that invitation without interference.

The Fatal Assumption: Gravity as Geometry

Einstein believed falling bodies weren't accelerating—they were following geodesics. And so, he replaced force with shape. But in doing so, he replaced one mystery with another.

- Why does mass bend space?
- What is space made of, that it can curve?
- How does curvature propagate?

These questions were never answered—because GR never asked them. It replaced force with geometry, but never asked what geometry is made of.

UED gives the answer: geometry is a perceptual outcome of ripple dynamics in SVE. It's not what's real—it's what we *see* when we misunderstand alignment as motion.

Ripple Insight

"Falling isn't being pulled. It's being aligned with a low-tension ripple path." — UED Principle of Passive Descent

When you fall, you are not moving through space. You are settling into resonance with the underlying field structure. You are drifting toward equilibrium, not being forced away from it.

Einstein gave us the courage to remove gravity from its Newtonian pedestal. UED gives us the framework to remove curvature from its geometric throne.

In the next section, we'll take on the final act in Einstein's theater—the bending of light. And we'll ask the same question again:

Is light really bending? Or is something else in the field realigning—and taking our perception with it?

The Eclipse and the Starlight – The Curved Light Illusion

When the sky confirmed Einstein—but not the full truth.

On May 29, 1919, the universe bowed to a prediction made in a notebook.

During a total solar eclipse, Sir Arthur Eddington and his team observed something astonishing: starlight from behind the Sun appeared displaced—as if it had bent around the solar mass to reach Earth. It was the moment Einstein's General Relativity stepped from blackboard to starlight, turning theory into legend.

According to Newtonian physics, light—having no mass—shouldn't respond to gravity. But Einstein had proposed otherwise: if mass curves spacetime, and if light follows the straightest path through that curvature (a geodesic), then light should appear to bend around massive objects like the Sun.

Eddington's photographs, blurry by today's standards, shook the foundations of science. They showed stars slightly out of place—by exactly the amount Einstein had predicted.

The verdict was clear: The Sun had bent light. Einstein was right. Newton was dethroned. And spacetime was now the ruler of reality.

Or so we thought.

What the Sky Showed—and What It Assumed

The experimental result was not in dispute: light from stars near the Sun's edge during the eclipse reached us from a slightly shifted position. But what exactly happened in the field between the star, the Sun, and the observer?

Einstein's interpretation was elegant:

- The Sun's mass bends spacetime.
- Light travels through this curvature.
- What appears to us as a bending of light is simply the natural trajectory of a beam in a curved geometry.

But this entire explanation rests on one crucial assumption:
That light travels, like an object, through spacetime.
UED says this is the fundamental misreading. The light never bent. Our perception of alignment shifted—and that illusion was interpreted as curvature.

UED Reinterpretation: The Ripple Refraction Principle

In the UED framework, light is not a traveling wave or particle. It is a ripple alignment event between a source and an observer across the Structured Vacuum Energy (SVE) field.

The Sun, being a massive body, thickens the ripple field around it—compressing the ripple density, like a lens thickening at its center. This doesn't bend space—it alters the perception alignment gradient across the field.

When a ripple from a distant star approaches the observer's field, it must traverse the Sun's dense ripple zone. This shifts the alignment trajectory, just like light appears to bend when it passes from air into water—not because it curves, but because its medium changes.

What Eddington observed was not the bending of a light beam. It was the refracting of ripple alignment, due to a dense ripple field near the Sun.

Not a Gravitational Lens—But a Ripple Lens

Einstein's model treats the Sun as a gravitational lens, warping space itself.

UED refines this: The Sun is not warping anything—it is realigning the ripple environment. The SVE field around it is denser, more reactive, and causes the ripple path to reconfigure before it reaches the observer.

The result? We perceive the light source to be shifted—but the light never changed direction. It didn't travel around the Sun. It aligned through a distorted medium.

Perception, Not Photon, Was Displaced

In GR, light bends *because space bends.*

In UED, nothing bends. The ripple from the source engages the observer's field via the path of least resistance, defined by ripple tension gradients.

That path is not curved—it is rerouted.

The photon didn't take a detour. Your mind's alignment with the field did.

This explains:

- Why we see the shift.
- Why it's consistent.
- And why the explanation doesn't require space to be anything other than a ripple medium.

Ripple Insight

"Mass doesn't curve the fabric of space. It thickens the ripple field, changing alignment trajectories." — UED Field Refraction Principle

Just as a denser medium like glass bends light through refraction, the SVE field becomes optically denser near large masses—not in the classical sense of optics, but in terms of field ripple density.

The star wasn't moved. The light wasn't bent. Your connection to it was altered by the Sun's ripple environment.

Why This Changes Everything

Eddington's experiment proved that something happens near massive bodies that affects how we see distant light.

General Relativity interpreted this as curvature.

UED explains this as consequence-based refraction in a structured medium—a field effect, not a geometric one.

And the implications are enormous:

- It eliminates the need for spacetime as a substance.
- It simplifies light behavior into ripple mechanics.
- It redefines gravitational lensing as alignment shift, not path distortion.

- It opens the door to understanding light behavior in quantum and cosmic contexts without contradiction.

Einstein's prediction was correct in magnitude—but not in mechanism. UED honors the observation, but rewrites the cause.

In the next section, we'll step back from these experiments and ask: What didn't Einstein claim? What caution did he offer that the world ignored? And why it matters that UED continues not just the theory—but the spirit of Einstein's curiosity.

What Einstein Never Claimed

The map was brilliant—but the territory remained unknown.

It is important—essential, in fact—to distinguish between Einstein's legacy and the mythology that followed him.

Einstein never claimed to have discovered the ultimate nature of space, time, or gravity. He never insisted that spacetime was a tangible thing—something physical that could stretch, bend, or ripple like rubber. He used geometry as a language, a mathematical framework for modeling how mass and energy appear to behave in the universe.

In his own words:

"Time and space are modes by which we think and not conditions in which we live." — Albert Einstein

And elsewhere:

"The true nature of space and time remains a mystery."

Einstein didn't mistake his metaphors for matter. He didn't worship curvature—he used it, cautiously, to explain measurable consequences like light deflection and gravitational time shifts. His goal was not to answer all questions, but to advance the conversation.

It was the world that elevated him to sainthood. It was the scientific culture that turned a model into a gospel, and a map into a dogma.

A Cautious Thinker, Not a Theoretical Prophet

Einstein was deeply humble about the limitations of his theory. He admitted frequently that General Relativity did not provide a full understanding of reality—especially at quantum scales or at singularities.

He distrusted blind formalism. He detested untested speculation. He remained uncomfortable with the fact that spacetime, in his model, seemed to act like a substance—yet had no defined composition or structure.

He never declared that space could bend like a material surface. He never claimed that time was a thing that flows or dilates independently. He never described light as a traveler moving through empty space. He modeled appearances—and left the ontology open-ended.

The Worship Came After Him—Not From Him

The idolatry of Einstein came later—not from Einstein, but from those who stopped where he paused.

His thought experiments became sacred stories. His equations became religious texts. Curvature became not just a symbol of beauty, but an unquestionable cause. As generations of physicists chased elegance over essence, the metaphors turned into material beliefs:

- That space was a literal fabric.
- That time could bend and stretch.
- That gravity emerged from geometry alone.
- That mathematical beauty was equivalent to truth.

Einstein never endorsed these dogmas. He invited exploration, not stagnation.

UED Doesn't Refute Einstein—It Completes the Climb

Universal Energy Dynamics does not set out to tear down Einstein's theory—it sets out to climb beyond it, to follow the very path Einstein himself started.

It respects the brilliance of his thought experiments, but asks the next logical question:
What is the ripple field beneath the geometric illusion?
UED takes the curvature Einstein described and reinterprets it as a perceptual outcome of ripple alignment in a structured medium. It accepts the mathematical elegance, but redefines the source.
It doesn't discard the train, the elevator, or the eclipse. It descends into their foundations—to reveal the hidden assumptions that shaped them.

- That light travels through space.
- That time is something that can dilate.
- That motion alters reality.
- That perception equals truth.

Each of these assumptions was baked into the thought experiments. They made the logic work—but they also silenced deeper questions.

Thought Experiments: Light and Shadow

Einstein's thought experiments were brilliant beams of light into the dark corners of physics. But like all light, they also cast shadows—the assumptions they did not illuminate.

UED returns to those shadows—not to dismiss the light, but to see what it missed.

Because a thought experiment is only as powerful as the questions it asks—and the questions it forgets to ask.

Closing Reflection

"Einstein didn't claim to have seen the whole truth. He only dared to look where others refused. But the idol we built was never him—it was the comfort of stopping the search."

The next chapter will begin stripping away those comforting assumptions. Not to dishonor Einstein—but to honor what he would have done next:

Ask the next impossible question.

Chapter 3: A Beautiful Mistake

What GR got right — and what it misunderstood about space, time, and perception.

"Some equations are so beautiful, we stop asking if they're true."

There's something seductive about elegance. When mathematics fits reality with such precision that it feels poetic, we forget to question what lies underneath. Einstein's General Relativity was one of those moments—a theory so graceful, so predictive, so aligned with observation that the world called it genius and declared the conversation over.

It wasn't.

General Relativity wasn't wrong. It was a stunning interpretation of how things behave. It predicted the precession of Mercury's orbit, explained gravitational time shifts, accounted for the bending of starlight, and even redefined our understanding of the cosmos. But like every interpretation, it was built on assumptions—and those assumptions were never dissected. They were canonized.

- That space is a real thing that curves.
- That time is a dimension that stretches.
- That light is a traveler moving through a vacuum.
- That perception is passive and geometry is truth.

None of these were proved. They were inherited beliefs, dressed in new language. And so GR became a masterpiece—a beautiful mistake that confused description for cause, map for territory, and consequence for origin.

This chapter is not a rejection. It is a reckoning.

We will explore what GR got brilliantly right—and what it misunderstood at its core. Not to diminish its legacy, but to complete it, by stepping into the space it never entered:

The substrate. The ripple. The perception-aligned field that makes the illusion of curvature possible.

Because behind the beauty, something even more fundamental was waiting to be seen. And it's time we stopped worshipping the elegance— and started understanding the engine.

The Accuracy of Illusions

When matching appearances became more important than understanding reality.

General Relativity is, without doubt, one of the most celebrated theories in the history of science. Even its critics must concede that it has produced a track record of empirical success that few theories can rival. From the subtle wobble in Mercury's orbit to the precise functioning of satellite navigation systems, GR doesn't just work — it performs with surgical precision.

Let's revisit the highlights:

- **The Precession of Mercury:** For centuries, Newtonian mechanics couldn't fully explain why Mercury's orbit shifts slightly with each revolution. GR corrected the discrepancy by treating gravity not as a force, but as the curvature of spacetime — and the numbers matched.
- **Gravitational Lensing:** Light from distant stars bends as it passes near massive bodies like galaxies or black holes. GR predicted this long before we observed it. And when we did? The deflection angles agreed almost exactly.
- **Gravitational Time Dilation:** Clocks in stronger

gravitational fields tick more slowly. Satellites in orbit must account for this effect — otherwise, GPS would drift by kilometers each day. GR's corrections are essential for your phone's location accuracy.

- **Black Hole Horizons & Frame Dragging:** From LIGO's detection of gravitational waves to NASA's Gravity Probe B measuring the warping and twisting of spacetime, GR has been repeatedly supported by observation.

On paper, the theory is a triumph.

And yet, beneath that triumph lies a critical mistake — not in the math, but in the meaning.

The Map Is Not the Terrain

There is a dangerous illusion in science: when equations predict observations, we assume they describe reality.

But prediction is not explanation. And mathematical precision is not metaphysical truth.

Einstein's equations model how mass and energy affect the motion of objects and the path of light. But they don't say what space is. They don't say what time is. They simply describe how these quantities must behave if we assume they are dimensions that bend.

That's a big if.

It's like using a flat Earth map to navigate — and getting where you need to go. The accuracy doesn't make the Earth flat. It just means your model works well enough within a limited context.

GR works not because it captures the essence of the universe, but because it models how phenomena appear under certain assumptions — namely, that space is a smooth continuum, that time is geometric, and that gravity is a distortion of these invisible properties.

UED doesn't challenge the success of GR's outputs. It challenges the origin of those outputs — and whether we've mistaken illusion for structure.

The Physics of Appearances

What if the bending of light isn't caused by curved spacetime, but by ripple alignment in a denser field? What if the slowing of clocks isn't due to warped geometry, but to delayed perception in ripple-saturated zones?

UED suggests that many of GR's predictions remain valid because ripple dynamics and field tension create the same effects, but from completely different causes.

The math overlaps not because spacetime is curved, but because our perceptions are ripple-bound, and ripple gradients simulate the same outcomes GR attributes to geometry.

In other words:

GR is a brilliant shadow play. It captures the silhouettes of ripple consequences — without ever seeing the ripples.

Framing Insight: Right Results, Wrong Reality

Einstein wasn't wrong about what happens. He was wrong about why it happens.

- He described the effect of ripple misalignment as spacetime curvature.
- He interpreted field delay as time dilation.
- He viewed ripple lensing as geodesic deflection.

Each observation was correct. Each interpretation wasn't the only one possible.

UED doesn't deny the data. It reframes the data through consequence mechanics—by returning to the medium Einstein never defined: Structured Vacuum Energy (SVE).

This is why GR became a beautiful mistake. Because it worked so well, no one dared ask what it left unexplained.

And yet, beneath the elegance, there remains a deeper truth:
Working models don't make a worldview complete. They make it convenient. But the universe is not obligated to be convenient. It is only obligated to be real.

In the next section, we'll dive into the three foundational misreadings that gave GR its form—so we can begin to understand where description ends, and reality begins.

The Three Big Misreadings

Where Relativity described the results but misunderstood the roots.

Every great theory leaves behind traces of the era it was born in. General Relativity (GR), for all its revolutionary power, still carried within it the conceptual fossils of classical thinking. It was bold enough to rewrite gravity as geometry — yet it never challenged the deeper assumptions that framed space, time, and light. Instead of questioning the ontology, GR upgraded the equations.

The result? A theory that matched observations, but misread the foundations of reality.

Here are the three most consequential misinterpretations at the heart of GR — and how Universal Energy Dynamics (UED) corrects them.

a) Space as Substance

The myth of the flexible void.

GR treats space as if it's a real, physical entity — a sort of invisible fabric that can bend, warp, ripple, and even tear under the presence of mass and energy. It speaks of spacetime curvature as if space were a kind of medium, pliable and responsive.

But ask a simple question:

What is space made of?

GR doesn't answer. It doesn't even attempt to. It simply assumes that space is a thing — and moves on.

There is no defined medium. No physical mechanism explains how space holds tension, how it curves, or how it responds dynamically to mass. Geometry becomes action. Coordinates become causal. But the substance of space is left undefined.

This leads to absurdities:

- If space is a fabric, what holds the fabric?
- If it curves, where does it curve *into*?
- What resists the bend?
- What carries the curvature across distance?

These questions are dismissed — because GR is too busy solving equations to ask what they're really describing.

UED Correction: Space Is Not a Thing — It's a Consequence

In UED, space is not an entity. It is not a fabric, substance, or container. It is the absence of resistance within the Structured Vacuum Energy (SVE) field — a background of stable, ripple-capable energy. When this field is undisturbed, there is no space, no motion, no distance — only stillness.

What we experience as "space" is actually the ripple structure that emerges when the SVE is disturbed by the presence of energy, mass, or intent.

Ripples form gradients of interaction, and those gradients define how energy and observers perceive *distance* and *location*.

So space is not curved.

It is shaped by ripple gradients.

And where GR sees curvature, UED sees field tension maps.

Space is not what bends.

It is what emerges when ripples try to realign.

b) Time as a Flexible Dimension

The illusion of stretching the unmeasurable.

GR tells us that time can bend, slow, or stretch depending on your velocity or gravitational environment. According to the theory, clocks near a massive object tick more slowly, and time for a moving object dilates as it approaches the speed of light.

It's all very elegant — until you ask:

What exactly is "time" that it can stretch?

Time, in GR, is treated as a fourth dimension, akin to space — measurable, traversable, deformable. But this is a conceptual shortcut, not a physical definition. GR never explains what time is. It just inserts it into the coordinate system and warps it like rubber.

But time is not a thing. You can't isolate it, bottle it, or detect it apart from change.

GR assumes time flows — but what flows? What carries time forward? What resists it? What stretches when time dilates?

Again, silence. Because GR is a model of effects, not of causes.

UED Correction: Time Is Not a Dimension — It Is Emergent

In UED, time is not a fabric or a dimension. It does not exist as an independent entity that flows. Rather, time is a consequence — the measurable delay between the cause (a ripple) and the recognition of its alignment in the observer's field.

This delay is called Alignment Threshold Time (ATT) — the smallest unit of time in which perception becomes possible. ATT depends on:

- The observer's ripple state.
- The ripple density of the surrounding SVE.
- The interference patterns between internal and external fields.

Thus:

- Time appears to slow near massive bodies because ATT increases due to denser ripple tension.
- Time appears to dilate at high speeds because relative misalignment delays ripple engagement.

Time doesn't stretch. Perception lags.

We don't experience "time passing." We experience ripples being processed by our field state — and the interval between ripple and resonance is what we label "time."

c) Light as a Traveling Entity

The great illusion of motion without a medium.

GR assumes that light moves — like a wave through space, or a photon particle zipping across the vacuum at constant speed. It treats light as a traveler in a spacetime landscape.

But this notion, too, is borrowed from pre-relativistic physics — and was never re-examined deeply.

Let's pause and ask:

Have we ever observed light *in motion*?

Not really. We observe only its point of emission and point of interaction. Everything in between is inferred — and assumed to be linear, consistent, and directional.

But there is no direct observation of light in transit. No one has ever "seen" a photon mid-flight. It is always either already there, or not yet arrived.

UED Correction: Light Doesn't Travel — It Aligns

Under UED, light is not a thing that moves. It is a ripple resonance — a field alignment event between two ripple zones: source and observer.

When alignment is achieved, illumination happens. The delay between emission and perception is not "light traveling"—it is the time required for ripple realignment across the SVE field, governed by the ATT and field interference.

This means:

- Light doesn't bend — the ripple path reconfigures across high-density zones (like near the Sun).
- Light isn't blocked by distance — it is delayed by alignment

effort through the field.

- Light isn't constant because of a universal speed limit — it appears constant due to uniform ATT under ideal conditions.

Light is not a traveler. It is a handshake between ripple fields — instantaneous from the frame of interaction.

Conclusion: Geometry is Not Ontology

GR's beauty lies in its equations. But equations are not existence. They describe how things *appear*, not what things *are*.

Each of GR's misreadings — of space, time, and light — comes from treating appearance as cause, and geometry as substance.

UED steps beyond these illusions by returning to ripple-first thinking. It shows that beneath the mathematics of curvature lies a field-based universe, shaped not by coordinates, but by consequence mechanics in Structured Vacuum Energy.

In the next section, we'll address the heart of it all — perception — and how the observer's ripple field plays the starring role that GR never acknowledged.

The Perception Problem

What happens is not what you see—what you see is what happens through you.

For all its brilliance, General Relativity (GR) is strangely silent on one of the most crucial players in any physical interaction: the observer.

GR models gravity, light, and time as phenomena that occur independently of perception. It assumes the observer as a passive coordinate reference, someone who simply reads a clock or sees a star, unaffected by their own field state.

This assumption is subtle—but it creates a fundamental blind spot in the theory:

It treats perception as neutral. UED shows perception is anything but.

Perception as Passive: The GR Assumption

GR defines observation in terms of reference frames. A frame moves, accelerates, or exists in a gravitational field, and physics adjusts accordingly. But the theory doesn't account for how the observer's internal field state interacts with the external environment.

In GR:

- A moving observer sees time dilate.
- A stationary observer sees light bend.
- A clock on Earth ticks differently than one in orbit.

But the observer is always an external witness to these changes—not a participant. The act of observing is modeled as measurement, not as ripple interaction.

This view, while mathematically convenient, ignores the energetic reality of perception—the fact that observation is not just passive reception but active field engagement.

UED Correction: Perception is Alignment

In Universal Energy Dynamics (UED), observation is not a neutral act. It is the final step in a chain of ripple interactions. The observer's own field—their structured energy body—must achieve a certain level of alignment with the incoming ripple field for perception to occur.

This is true for:

- **Light**: The ripple from a source must match the field configuration of the observer to be registered as a visual event. If the ripple doesn't align, nothing is seen—no matter how much energy is present.
- **Time**: The experience of time depends on how quickly the

observer's field can register change, i.e., how it processes ripple realignments. This delay is governed by ATT (Alignment Threshold Time).

- **Position**: Spatial awareness is a product of field interaction points, not static coordinates. Where something "is" depends on where its ripple intersects with yours.

Thus:

Observation is not measurement. It is resonance. Seeing, hearing, timing—all emerge from field-perceptual alignment.

Example: Rethinking Time Dilation

Let's return to one of GR's core concepts: time dilation.

In GR, time appears to slow down near massive objects or at high velocities. This is explained by curved spacetime affecting clock rates. But this model, again, assumes the clock operates independently of the observer's perception of it.

UED reframes this entirely.

Time does not slow. Your perception of time changes because the ripple environment changes.

A massive object like Earth creates a denser ripple field, meaning the observer's internal field takes longer to achieve resonance with incoming ripple patterns. The delay is not in time itself—it is in perception latency due to increased ripple resistance.

Similarly, a fast-moving object disrupts its own ripple coherence with the external field, making the ATT longer. The result? Events are perceived as happening more slowly—not because "time stretches," but because field alignment takes longer to stabilize.

A clock doesn't slow because it's moving. You perceive it as slow because the field alignment delays its ripple echo to your awareness.

Key Insight: Perception is the Final Consequence

All physics begins with energy, evolves through structure, and ends in perception. But GR stops at structure—and skips the last, most human, most experiential stage.

That's where UED begins its expansion.

The observer is not a bystander. They are a field in interaction, and everything they observe is filtered through that field's alignment status.

- If your field is tense, you perceive slower.
- If your field is coherent, you perceive clearly.
- If your field is distorted, you miss the signal entirely.

This isn't philosophy. It's energetic mechanics.

"Perception is not neutral. It is the final ripple in a chain of field interactions." — UED Perceptual Principle

Why GR Had to Miss This

Einstein never integrated the observer's field state because GR is macroscopic and geometric. It assumes a clean world of ideal observers and external curvature. But in the real world, there is no separation between the observer and the observed. Every act of seeing, feeling, and measuring is a ripple event.

UED restores that connection. It shows that space, time, and light are not "out there" to be seen—they are formed at the edge of your perception, built in real-time as your field meets the ripple landscape.

This reframing brings us to the final conclusion of this chapter: that even the most precise theory can become an illusion if it forgets the lens through which all things are known—the perceiver.

In the next section, we'll explore why GR's beauty became its blinder, and why the path forward must begin beneath the curvature, in the ripple field it never acknowledged.

A Mistake Worth Making

When a brilliant model blinded us to a deeper truth.

Not all mistakes are failures. Some of them are landmarks—turning points where human thought evolved by stretching the imagination into unfamiliar territory. General Relativity (GR) was one such moment.

It was a beautiful mistake.

A theory that dared to bend the structure of space, stretch the meaning of time, and reframe gravity not as a force but as the architecture of reality. It took the messiness of classical mechanics, the paradoxes of Newtonian gravity, and the failures of flat models—and replaced them with an elegant, predictive geometry.

Einstein gave the universe a script. It was cohesive. It was breathtaking. It worked. And it was incomplete.

When Models Become Monuments

GR was never meant to explain everything. Einstein himself knew this. Yet the scientific community, mesmerized by the success of the equations and the sheer poetic elegance of curvature, turned this model into a monument. Not a stepping stone—but a stopping point.

We built a palace on top of curved spacetime and forgot to check the ground beneath it.

We thought we had discovered the floor of reality—when we had only mapped its shadow.

GR's greatest limitation is not its math. It's its ontology—its silence about what space, time, and light actually *are*. By turning geometry into cause, it placed abstraction above substance. And in doing so, it buried the true energetic mechanics of the universe under a layer of mathematical metaphor.

What could have been a transitional theory became a terminus. The floor was never questioned, because the stage looked so perfect.

Beauty as Barrier

There's a reason beautiful theories are dangerous. Their elegance discourages dissent. When something works so well, we stop asking whether it's built on sand.

- We admired how mass "curved" spacetime—but we didn't ask *what was curving*.
- We modeled time dilation to the last decimal—but forgot to ask *what time is*.
- We watched light "bend" around stars—but never questioned *how light moves through nothing*.

In every case, a working equation covered a missing cause.

The deeper truth? We weren't bending space. We were watching ripples realign in a structured field that GR never saw.

The Hidden Layer Beneath the Geometry

UED doesn't replace GR. It reaches below it.

It doesn't argue with the predictions—it reframes the premises.

- Where GR sees curvature, UED sees ripple tension.
- Where GR bends spacetime, UED shows how ripple gradients create perceived paths.
- Where GR treats time as dimension, UED explains it as alignment latency in perception.
- Where GR draws motion with coordinates, UED reveals movement as realignment in SVE.

The difference is not the outcome—it's the origin. UED explains what GR could not: why things appear as they do.

And that "why" changes everything.

A Mistake Worth Making

Let us say it clearly: Einstein's curvature was not a failure—it was a brilliant detour. It simplified a chaotic world and gave us the courage to challenge Newton's invisible hand. But it did so by trading one mystery for another—and building elegance on assumptions.

The tragedy is not that GR made a mistake. The tragedy is that we worshipped the mistake instead of outgrowing it. And now, over a century later, the opportunity is not to destroy GR—but to complete it.

To return to the floor. To look beneath the stage. To see the Structured Vacuum Energy field, rippling silently beneath our models, shaping reality not through curvature, but through consequence.

"Curvature explained the stage. But UED shows you the floor beneath the stage—The field that holds everything together."

Chapter 4: When Assumptions Become Law

How static models, coordinate dependence, and belief in absolutes shaped science

"The most dangerous assumptions are the ones no one remembers making."

Before Einstein curved space, Newton boxed it. Before we measured the speed of light, we decided it must be constant. And long before time became elastic in relativity, it was already mistaken for something real.

Modern physics, for all its brilliance, stands on quiet foundations—assumptions so deeply embedded in the scaffolding of thought that we no longer see them. We don't test them. We build cathedrals of theory upon them.

This chapter is about those forgotten choices—those unquestioned defaults that turned into laws. Not because nature confirmed them, but because no one challenged them long enough to see beyond.

We will expose the cage of coordinates, where real motion became a projection of reference frames. We will unravel the illusion of continuous time, born not from experience but from calculus. And we'll confront the enshrinement of absolutes—like the speed of light—as if they were sacred relics rather than ripple-dependent readings.

The truth is: the universe was never governed by equations. It was animated by dynamics—by structured fields, ripple flows, and observer alignments.

But assumptions, once they harden into belief, become invisible. They form the floor of science's mental theater—and any stage built upon them reflects only part of the play.

In this chapter, we tear up that old flooring. We go looking for the assumptions no one dared question—not to discard our models, but to reveal the terrain beneath them.

It's not about proving GR or Newton wrong. It's about asking: *What if we began before the coordinates? Before the curves? Before the constants?*

What if we started with field, not frame? Ripple, not reference? Consequence, not coordinate?

The next sections begin this excavation.

The Coordinate Cage – Static Maps of a Dynamic Field

"A map is not the terrain. But we've spent centuries mistaking coordinates for reality."

At the heart of modern physics lies a deeply entrenched illusion — that the universe can be fully described by coordinates. From Newton's inertial grids to Einstein's relativistic frames, science has long clung to the idea that space can be divided, labeled, and measured into tidy boxes. Reality, it was assumed, must submit to the graph paper.

But what if the universe never followed a grid?

The Rise of the Grid Mentality

The scientific revolution gave birth to mathematical modeling. With it came Cartesian coordinates — a revolutionary way to describe position, motion, and change. For Isaac Newton, this was a godsend. He introduced absolute space and absolute time: space was like a great invisible container, and time ticked uniformly like a divine metronome. Every object had a position, velocity, and acceleration — all definable within this sacred box.

But while the math worked, the assumption hardened: that location, motion, and interaction could all be captured by reference to a coordinate system. Over time, this wasn't just a method — it became a worldview.

And then came Einstein.

Einstein's Frame Shift — But Still in a Cage

Einstein radically challenged Newton's absolutes by introducing relativity: space and time weren't fixed — they depended on the observer's motion. Yet, ironically, Einstein's genius still relied on coordinates — just more dynamic ones.

Where Newton gave us a static grid, Einstein offered flexible coordinate frames, curved and distorted by mass and motion. But the trap remained: we were still in the cage of coordinates. The assumption endured — that physical reality could be faithfully mapped by mathematical locations, regardless of what was actually happening beneath.

The obsession continued. Every theory that followed — from quantum field theory to string theory — sought to locate particles, waves, energies *somewhere* in a field of x, y, z, and t.

But what if location is not a thing to be measured — but a phenomenon to be understood?

The UED Correction: Alignment, Not Address

In Universal Energy Dynamics (UED), we discard the cage. We no longer treat space as a container or time as a clock.

Instead, we view the universe as a structured ripple field — always active, never still. Nothing has a fixed location because nothing is at rest. What we experience as location is the consequence of ripple alignment between an observer and an external field event.

- You don't see the Sun "at a point in space" — you perceive the ripple that aligns with your observation field.
- A star a million light years away isn't "there" — it's perceived now, through the ripple tension that allows your field to receive it.

- Even the chair beneath you doesn't occupy a coordinate — it maintains a real-time field engagement with your body's ripple signature.

From this view, coordinates are like street addresses in a dream — they feel useful until you wake up.

UED shows us that the very idea of fixed position is an artifact of our modeling, not a property of the universe. What's real isn't the map — it's the flow of energy, the tension gradients, the resonant interactions within a constantly shifting field.

This also explains why GPS works even though the Earth isn't truly still — because it syncs signals, not absolute positions. It measures ripple synchronization, not universal truths of place.

Escape from the Coordinate Cage

The consequence of clinging to coordinates is profound:

- We miss the dynamic nature of energy in motion.
- We create paradoxes (like superposition and entanglement) because we expect particles to *be* somewhere.
- We over-engineer models to fit static grids, while the universe flows like a field in consequence.

By releasing ourselves from the idea that "to exist" means "to be located," we step into a more fluid, accurate, and ontologically sound description of reality.

In UED, everything is a relation. Nothing stands still. Nothing "occupies" — things only interact.

Key Insight:

"Coordinates describe what is convenient, not what is real. What is real is ripple alignment — not location, but engagement."

The Illusion of Continuity – Why Static Time Fails

"Time does not flow. It flickers into perception—one alignment at a time."

From sundials to atomic clocks, from ancient myths to Einstein's spacetime, humanity has long believed in the flow of time—as if it were a river that carries all events downstream. We say time "moves," "passes," or "flies." In physics, we represent it as a line, a dimension, a coordinate "t." Smooth. Continuous. Absolute in Newton's age, relative in Einstein's—but always there, flowing beneath everything.

But what if nothing ever flowed?

What if time, as we perceive it, is not a substance, not a background stream—but a consequence of interaction, a punctuated realization born at the moment of field alignment?

This is the core illusion UED dismantles. And it starts with reexamining one of the most revered assumptions in all of physics: that time exists independently of perception.

The Continuous Time Assumption: Legacy of the Line

Classical physics and calculus together painted a compelling picture: that all events unfold in an uninterrupted continuum. Using tools like dt (an infinitesimal unit of time), scientists could model everything from the motion of planets to the vibrations of atoms.

This calculus-based framework made the world predictable—and smooth.

But it also introduced a belief: That time is infinitely divisible, and that reality must obey this mathematical smoothness.

From Newton's flowing time to Einstein's curved time, this continuity was never questioned. It simply became the canvas for all theories.

Yet in experience, time doesn't feel continuous. We don't live in a stream—we live in moments.

Realizations. Decisions. Blinks. Shifts. Memory itself is punctuated, not smeared.

So where does this illusion of smoothness come from?

The Trap of the Infinite Slice

When we use dt or dx, we're not touching reality—we're describing it with assumed smoothness. The mathematics assumes continuity so it can calculate change. But nature may not operate on that assumption.

This creates a paradox: Calculus helps us describe systems, but in doing so, it imposes a structure—a model of continuity—that may not exist in the raw field of the universe.

As a result, physics built on continuous time (e.g., Schrödinger equations, spacetime intervals, Lagrangian mechanics) may match outcomes but misrepresent what's fundamentally happening.

We model ripples as smooth sine waves—yet perception occurs in bursts, not flows. We model motion as change across time—yet what moves is not the object, but the ripple engagement.

UED's Correction: Time is a Consequence, Not a Container

Universal Energy Dynamics (UED) reframes time not as a flowing dimension, but as a perceived interval that arises only when ripple alignment occurs. Specifically, it emerges from the ATT — Alignment Threshold Time — the minimum delay between an observer's intent field and the interaction of external ripples.

In this view:

- Nothing is happening "in" time.
- Events are not placed along a timeline.
- Instead, each perceived moment is a node of alignment, where field tension permits interaction.

Time is not the backdrop. It is the consequence.

When ripple fields interact—when tension gradients permit resonance—a moment becomes visible. If no alignment occurs, there is no time experienced, no change observed, no memory created.

This explains:

- Why time "slows down" in fear (high ripple tension causes denser perception slices).
- Why time "disappears" in flow states (stable alignment minimizes ripple friction).
- Why trauma loops feel endless (misalignment recycles ripple delay, distorting ATT).

Implications: The "t" in Physics Has an Ontological Gap

From Newton to quantum mechanics, every equation relies on t—a parameter treated as real, reliable, and foundational. But what is t?

No theory truly defines it. No experiment isolates it. It is always implied. Assumed. Needed for the math, but absent from the medium.

UED forces a fundamental reconsideration:

- Time is not a measure. It is a product of perception.
- No ripple = no ATT = no time.
- Different alignments produce different time experiences, not because clocks are wrong, but because reality is local to perception.

Thus, all physics built on "t" must face the truth: you're not measuring time—you're measuring ripple delay.

Key Insight:

"Time isn't flowing through you. You're flickering through alignment."

The continuity of time is a model convenience—not a universal truth. In reality, nothing continues unless the field allows it to align.

The Myth of Absolutes – Constants, Invariants, and Scientific Dogma

"What we call constants are just the outcomes of repeated alignments—not the fingerprints of God."

In the modern scientific imagination, a few sacred symbols reign supreme. c, the speed of light. G, the gravitational constant. h, Planck's constant.

These are the pillars upon which entire disciplines are built—relativity, quantum mechanics, cosmology. They are not questioned. They are measured, refined, tabulated to decimal precision... and then treated as absolute truths. Immutable. Universal. Divine.

But what if they were never meant to be worshipped?

What if these so-called "constants" are not cosmic commandments—but contextual readings, specific to how we align with the field at a given point?

This section breaks that illusion. Because once constants become unquestionable, science stops asking better questions.

Constants as Frozen Consequences

Let's begin with a blunt truth: every physical constant in science is measured, not revealed. We observe, calculate, and calibrate these numbers from specific conditions—and then assume they hold across all scales and situations.

But UED argues: they don't.

In Universal Energy Dynamics (UED), constants are not foundational laws. They are reliable ripple effects—consistent only under specific field alignment patterns.

Take c, the speed of light in a vacuum. We treat it as an upper limit, a boundary of causality, a "speed of truth." But no one has ever "seen" light travel at 299,792,458 m/s. We detect ripples aligning at that rate under certain field conditions—and then infer motion.

From UED's lens:

- c is not a speed; it is the threshold delay (ATT) at which ripple information aligns with an observer's field.
- In high-tension environments (e.g., near black holes), c

appears to bend—not because it changes, but because the ripple conditions warp.
- In quantum entanglement, we observe instantaneous correlation—which breaks the limit of c, unless we accept c as a contextual limit, not an absolute one.

The same goes for G and h:

- G doesn't measure "gravitational strength" in empty space—it quantifies field tension gradients around massive bodies.
- h, Planck's constant, doesn't encode quantum reality—it represents the minimum ripple threshold for perceptual exchange.

None of these are the laws themselves. They're results—reliable, yes—but always field-dependent.

The Consequence of Absolutism

When constants become sacred, science loses agility. Instead of questioning the conditions that give rise to these values, we begin building entire frameworks assuming they are always true.

This is how we ended up with:

- Paradoxes in simultaneity (Einstein's train experiment).
- Causal breakdowns in quantum entanglement.
- Mathematical monstrosities like renormalization in quantum field theory.
- A string theory landscape of 10^{500} untestable universes—all clinging to a scaffold of untouchable constants.

Instead of adapting to a fluid, ripple-driven universe, science got stuck defending static scaffolding. And worse: models that didn't conform—like variable light speed, sub-Planck phenomena, or ontological frameworks like UED—were dismissed not on merit, but on noncompliance with sacred constants.

UED's Ripple Redefinition

In UED, all constants are seen for what they are: Ripple outcomes of structured vacuum tension, observed from specific field alignments.

There is no ultimate "speed of light." There is only the minimum ATT threshold for ripple perception under vacuum tension X.

There is no universal gravitational constant. There is only the average ripple gradient response in structured vacuum near mass concentrations.

Even Planck's constant is not fundamental—it marks a threshold of engagement, not an eternal law.

Thus, UED treats all constants as ephemeral but repeatable results, not rules of the universe.

■ Rethinking Scientific Dogma

If we release our grip on constants as cosmic commandments, we open the door to a more adaptive, truthful physics. One that:

- Reflects the universe's contextual intelligence, not rigid programming.
- Accounts for why constants appear stable (because ripple alignment is consistent under stable conditions).
- Explains why they sometimes don't (as in near-black-hole environments, entanglement, or early universe anomalies).
- Frees us from theoretical dead ends like "fine-tuning," "cosmic coincidences," and unprovable infinities.

Key Insight:

"Constants are not divine numbers. They are ripple responses—symptoms of alignment, not signs of design."

The real law isn't in the number. It's in the field that produced it.

The Rise of Belief Systems in Science

"Science begins with curiosity—but it can end in conviction."

Somewhere along the arc of modern physics, a dangerous transformation occurred.

Theories became frameworks. Frameworks became dogma. And dogma became law.

The transition was so gradual—and the mathematics so compelling—that few noticed when science stopped being an open inquiry into nature and became a cathedral of belief systems cloaked in equations. In this section, we unpack how elegant ideas hardened into unchallengeable truths and how Universal Energy Dynamics (UED) offers a necessary disruption—not by contradiction, but by consequence.

From Elegant Assumption to Institutional Axiom

Einstein didn't present curved spacetime as a religion. He proposed it as a model—a brilliant, elegant one—to interpret gravitational effects. Similarly, quantum mechanics wasn't born with metaphysical certainty. It was a tool to predict phenomena, not explain them ontologically.

But over time, elegance turned into enforcement.

What began as tentative descriptions of appearances—curved spacetime, probabilistic particles, dimensional strings—became assertions about reality itself. The very assumptions that launched revolutions in thinking were elevated to absolute truths, often without ever asking:

- What exactly is space?
- What exactly is time?
- What waves in a wavefunction?
- What bends when we say "spacetime curves"?

No answers—only metaphors. But metaphors repeated long enough become memorized truths. And memorized truths repeated across generations become dogmas.

Institutions as Echo Chambers of Consensus

Modern science operates within powerful structures of belief enforcement. These include:

- Peer-reviewed journals, which, while necessary for quality control, often reject ideas that question foundational assumptions.
- University curricula, which teach models as gospel, rarely encouraging students to investigate ontology.
- Public science communication, which simplifies theories for mass understanding but presents them as ultimate truths—"Space curves! Time dilates! Atoms are mostly empty!"

All of this forms a closed loop of validation. If your theory conforms to GR (General Relativity) or QM (Quantum Mechanics), you're welcomed. If it deviates—even with rigor, logic, and data—it's seen as "fringe," "philosophical," or "untestable."

Science becomes a tribe. And its truths become beliefs, not discoveries.

UED's Challenge: Re-centering the Observer

Universal Energy Dynamics (UED) doesn't just critique the math. It critiques the assumptions beneath the math—especially the omission of the observer's field from all physical equations.

In UED:

- Perception is not passive—it is a ripple alignment event.
- Observation is not external—it reshapes the local field.
- Constants are not sacred—they are outcomes of field conditions.
- Time and space are not containers—they are consequences.

Any framework that ignores the ripple field of the observer—or treats it as irrelevant—is ontologically incomplete, no matter how well it predicts outcomes.

This is the ripple insight GR and QM missed:

"An equation that leaves out the observer's field is a model of illusion—not of reality."

Why This Rigidity Matters

When science refuses to evolve its assumptions, it creates:

- Paradoxes it cannot solve (wave-particle duality, black hole information paradox).
- Mathematical scaffolding without physical grounding (string theory, inflationary cosmology).
- Blind spots in fields like consciousness, perception, and non-locality.
- Technological stagnation in areas where deeper understanding of the field could unlock innovation (energy transfer, propulsion, communication).

Belief is efficient. But only inquiry leads to truth.

UED reopens that inquiry—not by dismissing existing science, but by extending it into the substrate of the field where perception, reality, and ripple interaction truly originate.

Final Insight:

"Belief makes science safe. But truth makes it dangerous again."

When elegant assumptions go unchallenged, reality gets buried beneath equations. UED unearths it—not by rebellion, but by realignment.

Closing Thought: Law Is Not Truth

"The moment we confuse assumption with truth, we stop looking beneath."

Scientific progress is rarely stalled by ignorance. It is stalled by unquestioned confidence—by mistaking working models for reality, and elegance for essence.

Laws, however well-formulated, are not truths. They are interpretations. They are attempts to draw a boundary around appearances, to tame nature into equations that fit our current tools and thought structures.

But when a model becomes canon... when an assumption becomes untouchable... We begin to defend the map, rather than explore the terrain. We begin to polish the mathematics, rather than confront the ontological void beneath them.

This chapter has exposed how physics was shaped by unspoken defaults:

- That coordinates reflect reality.
- That time flows like a river.
- That constants are divine.
- That observation is neutral.

Each assumption served its time. But now, they block the door to a deeper physics—a physics that begins not with geometry, but with energy. Not with measurement, but with the field that makes measurement possible.

Universal Energy Dynamics invites us to tear up the flooring laid by outdated belief—and walk barefoot again on the raw energetic field beneath all phenomena.

Not to reject Einstein, Newton, or Bohr. But to continue the journey they began—past abstraction, into alignment.

A Shift in Direction:

We've spent chapters mapping space—curving it, measuring it, fragmenting it into coordinates.

Now we ask the deeper question:

What *is* space? Is it really empty? Or is it the silent body of something more intelligent?

The floor beneath curvature is about to rise.

Part II: The Ripple Reality

Chapter 5: The Structured Vacuum

Introducing SVE — Why the Void Isn't Empty, and the Universe Isn't Built on Space

We've spent centuries staring into space and calling it *nothing*.

Science has mapped the cosmos with telescopes, equations, and coordinates—yet it still clings to a primitive idea: that the universe floats in an empty container called "space." Even quantum mechanics, with its vibrating fields and ghost particles, treats the vacuum as a fuzzy zero—a stage for action, but not the actor itself.

But what if that's the biggest illusion of all?

What if the vacuum isn't empty? What if it's the most structured, intelligent, and foundational layer of all reality?

In this chapter, we introduce Structured Vacuum Energy (SVE)—the core of Universal Energy Dynamics (UED). Not a space to be curved. Not a void to be ignored. But the energetic floor of existence from which all ripples emerge and all phenomena unfold.

This is not a revival of the ether. It's not a metaphor. It's a consequential reality that explains what moves when nothing moves. What exists before mass. What connects observer and observed.

Einstein removed the ether to simplify the math. But he left behind a vacuum without substance—a map with no terrain. UED fills that missing layer with SVE: a still field that holds the potential for all motion, matter, time, and perception.

Here, the laws of nature aren't imposed—they are emergent ripple patterns. Here, light doesn't travel—it aligns. And here, gravity is not a pull—it's a slide down ripple slopes in a field that never moves.

This chapter reclaims the void—and gives it structure. Because before you can understand the universe, you must understand the ground it stands on.

The Myth of Emptiness

For centuries, scientists wrestled with a simple yet profound question: What lies between things? Is the universe filled with something—or truly nothing?

The Greeks called it *aether*—a mystical fifth element that filled the heavens. Newton believed in an absolute space, invisible yet real, a divine scaffolding on which the universe rested. Later, 19th-century physics gave form to the idea through the luminiferous ether, a medium believed to carry light waves just as air carries sound.

But in 1887, Michelson and Morley's now-famous experiment—designed to detect the Earth's motion through this ether—returned a puzzling result: no detectable drift. Light, it seemed, didn't care whether you were moving toward or away from it. The ether, if it existed, was silent.

Science panicked. And rather than question its assumptions about light and measurement, it declared the medium dead. The ether was dismissed as fiction, and with it, the very idea of a substantial vacuum.

From Substance to Stage

Einstein capitalized on this shift. He bypassed the need for a medium altogether by proposing that light's speed was constant in all frames of reference. His space wasn't filled with anything—it was geometry itself. Curved, elegant, measurable. But ultimately, empty.

The result? For over a century, mainstream physics treated space not as a thing, but as a container. A volume with no content. A placeholder. A silent background.

Even as quantum theory advanced and introduced the idea of a "quantum vacuum," the picture remained fuzzy. Physicists described the vacuum as seething with virtual particles and energy fluctuations.

But these were statistical ghosts—not structure. The vacuum was noisy, yes—but still thought of as nothing.

The UED Shift: Emptiness Isn't Empty

Universal Energy Dynamics (UED) challenges this dogma at its root.

The void is not absence. It is structure without resistance. It is the perfect alignment of potential—a state so harmonized that it offers no opposition, no signal, no reaction unless disturbed.

In UED, this foundational layer is called Structured Vacuum Energy (SVE). It is the *ground state of reality*, not a passive emptiness but a dynamic, ripple-capable field. It doesn't flow or vibrate unless triggered. It holds no particles, yet it contains the potential for all.

Why We Mistook It for Nothing

Emptiness, in our perception, is a lack of interaction. When we touch air, we feel nothing—so we say it's empty. But we're simply too coarse in our tools to detect the subtleties.

Likewise, when a region of the universe produces no resistance, no signal, no force—we label it "vacuum." But in truth, that stillness is not nothing. It is balance. It is a perfectly aligned state in which no ripple is yet born.

This leads to the core UED claim:

Emptiness is not the absence of substance—it is the absence of misalignment.

What we call "nothing" is actually everything in harmony.

Reframing the Vacuum

Thus, the myth of emptiness collapses. Space is not a void—it is a consequence of how ripples aren't aligning in a particular zone. And the vacuum is not an accidental leftover—it is the canvas of existence, designed to respond only when reality calls it to ripple.

UED returns physics to what it quietly lost after Einstein: A medium. But not an ether of particles. Not a fabric to stretch or tear. A field of intelligent potential, silently awaiting the spark of interaction.

Defining Structured Vacuum Energy (SVE)

Why the universe isn't built on space, but on equilibrium waiting to ripple.

What holds up the universe?

Not atoms. Not particles. Not space. Not even energy in motion.

Beneath everything we see, measure, or calculate lies something far more foundational—Structured Vacuum Energy (SVE).

Not to be confused with classical "ether" or quantum foam, SVE is not a substance that fills space, nor a fabric that can be curved. It is space-as-energy-in-equilibrium—a fully structured field that does not flow, move, or fluctuate until a ripple disturbs it. In that sense, SVE is not the *medium* of energy—it is energy, but in its most harmonized and undetectable state.

The Background Field of All Existence

SVE is the energetic base layer of reality. It doesn't emit anything on its own. It doesn't radiate. But it holds the capacity for all motion, mass, charge, and time to emerge. Unlike quantum vacuum which seethes with spontaneous fluctuations, SVE remains perfectly stable unless provoked.

You cannot detect SVE directly, just as you cannot hear silence. But you can detect its responses—when ripples form, propagate, align, or collapse. Every field interaction, every particle behavior, every gravitational or electromagnetic phenomenon is simply an expression of ripple dynamics within SVE.

It Doesn't Flow — It Holds Potential

Think of SVE as a still ocean, not an empty void. It's not flowing like wind or plasma. It doesn't need a directional force. It holds potential ripple states—waiting for an interaction, an intent, a consequence, to trigger motion.

Where mainstream physics focuses on the ripple, UED draws attention to the sea.

This is a crucial reversal:

- Modern theories start with what moves.
- UED starts with what enables movement at all.

And what enables it is structured harmony—SVE.

What SVE Enables

SVE isn't just a placeholder—it's an active foundation with unique properties:

- **Ripple formations:** All energy expression—whether light, gravity, or emotion—is a disturbance of equilibrium in SVE. Ripples are not objects; they are realignments of structured tension.
- **Field tension zones:** What we call *mass*, *charge*, or *force fields* are zones of condensed ripple pressure within SVE. These aren't separate "particles" or "forces" added on top—they are simply different levels of alignment tension.
- **Alignment Threshold Time (ATT):** The time we perceive is not clock-based—it is a consequence of how long it takes for our field to realign with ripple shifts in SVE. ATT explains the illusion of time delay, simultaneity distortion, and even time dilation.

Not a Medium, Not a Carrier

Here's the radical departure from 20th-century physics: SVE is not a medium through which light or energy moves. It is the condition for those expressions to emerge.

Light doesn't "travel" through SVE like a wave in water. Instead, light is the result of a successful ripple engagement between two field-aligned points in the observer's field and the source.

And unlike spacetime in General Relativity, SVE does not bend or stretch. It holds ripple gradients—fields of varying tension—that give the illusion of motion, pull, or curvature.

SVE: Energy in Equilibrium

This gives us the core UED claim:

Structured Vacuum Energy is not a substance in motion. It is energy at rest, perfectly balanced—until reality needs to happen.

Everything else—gravity, light, particles, consciousness—arises as structured distortions in this bedrock equilibrium.

Thus, the universe isn't built on space or time. It's built on a silent architecture of intelligent tension—a dynamic stillness that waits, listens, and ripples when called.

This is the return to substance without substance, to reality without reaction, until the moment of observation.

Why Nothing Moves — But Everything Aligns

Reality is not made of motion. It is made of ripple alignment.

What if the universe never moved?

What if everything we thought was in motion—falling apples, racing photons, orbiting planets, migrating galaxies—wasn't *traveling* at all?

Welcome to one of the most radical implications of Universal Energy Dynamics (UED):

Nothing truly moves. Reality is realignment.

The Great Misinterpretation: Motion

In classical physics, motion is defined as the change in position over time. Newton gave us this in terms of velocity and acceleration. Einstein refined it, embedding motion into curved spacetime where gravity "guides" the paths of things along geodesics.

But both models rely on the idea that objects move through space.

UED proposes something fundamentally different.

Space is not a stage. Light is not a traveler. Planets are not swinging around the Sun on invisible strings.

Instead, all entities are nodes of ripple tension within a structured field (SVE). What we interpret as movement is simply the reconfiguration of ripple alignment—the way tension balances and rebalances across the field.

Motion = Shifting Alignment

Let's reframe motion in UED terms:

- An object doesn't go from point A to point B.
- Instead, its ripple signature realigns with a different sector of the structured vacuum.

It's not the object moving—it's the field updating its state of tension.

The *observer* perceives this as motion, because the ripple pattern they interact with has shifted.

The experience of movement is a consequence of how field alignments evolve.

You don't see something move. You perceive a ripple shift.

Redefining Core Dynamics

This rethinking forces us to reinterpret some of the most fundamental forces in physics:

Gravity

Not a pull. Not a curvature. Gravity is simply a gradient in ripple tension across SVE. When two zones of different alignment tension interact, the ripple pathways naturally curve inward—but only because they seek tension equilibrium.
No force acts. The field simply responds.

Inertia

Traditionally defined as resistance to motion. In UED, it's the resistance of a field node to realign its ripple structure. A stable ripple node doesn't "want" to change state unless acted upon by a stronger ripple pressure.

Mass

No need for particles or Higgs fields. Mass is localized ripple density—a compressed tension pattern in SVE. Heavier mass = tighter ripple knots = higher resistance to alignment shifts.
Analogy: The Still Lake
Imagine a perfectly still lake. Now throw a pebble.
Ripples spread. Patterns shift. Reflections wobble.
To the uninformed eye, it looks like something is moving—crossing the water, drifting across the surface.
But in truth?
Nothing leaves the lake. Everything remains in place.
All that happens is energy redistributing through pattern changes.
This is UED's view of the universe:

- Structured Vacuum Energy is the still lake.
- Ripples are all reality.

- Motion is an illusion created by ripple alignment transitions.

Why the Illusion Works

Our perception is tethered to ripple feedback. Our minds interpret alignment changes as velocity, location, speed, and direction.

But underneath it all, no particle is hopping through space. There is only realignment—and the memory of prior alignments.

That memory gives rise to what we call motion. But in the deepest reality?

Everything is still, but nothing is stuck.

Ripple Insight

"Motion is not how things go from one place to another. It is how the field shifts to let things appear elsewhere."

This is the ontological upgrade UED offers: Not a denial of physics—but a correction of the substrate beneath it.

Where GR curves the map, UED realigns the field.

Where Newton spoke of forces, UED speaks of consequence-bound gradients in intelligent equilibrium.

And where Einstein saw light taking a curved path through spacetime, UED sees ripple engagement adjusting through field tension—instantaneous, emergent, and beautifully consequence-driven.

Revisiting Einstein — The Missing Medium

General Relativity removed the ether. UED brings back the floor.

Einstein gave us many gifts—mathematical elegance, conceptual breakthroughs, and the courage to question Newton's invisible forces. But in doing so, he removed something that had long underpinned physical thinking: the ether—the idea that the universe must have a medium through which interactions occur.

To be fair, Einstein was careful. He did not mock the ether concept outright. In fact, by 1920, he stated:

"According to the general theory of relativity, space is endowed with physical qualities... Therefore, there exists an ether; only we must give up the idea of its motion."

Still, the operational ether—a structured substrate with properties—was never formally integrated into his field equations. The idea was left dangling—neither alive nor dead. And that ambiguity has haunted physics ever since.

Curvature Without a Floor

General Relativity modeled gravity not as a force, but as the curvature of spacetime. Mass tells spacetime how to curve, and spacetime tells mass how to move.

Beautiful. Revolutionary. But one question was never answered:

What is spacetime made of?

In GR, spacetime is not a thing. It has no atomic structure, no field tension, no substrate. It is treated as a geometric stage—not a physical player.

Curvature occurs... in nothing. Gravity acts... through nothing.

This is where GR began to drift from ontological grounding into elegant abstraction.

The Carrier Einstein Never Defined

Think of it this way:

- A sound wave needs air.
- A water wave needs water.
- A ripple needs a surface.

What does a gravitational ripple need?

In Einstein's formulation, nothing. But in reality, something must ripple—or else the ripple is a mathematical hallucination.

UED steps in here with a clear, physically grounded answer:

Structured Vacuum Energy (SVE) is the missing medium. It is not a throwback to 19th-century ether—but a refined, consequence-aware, equilibrium-based field.

It doesn't flow. It doesn't vibrate like a string. It holds tension gradients—which we perceive as motion, force, mass, and yes... gravity.

Reframing GR's Effects

In GR, mass curves spacetime. That curvature alters the trajectory of particles and light.

In UED, mass is a condensation of ripple tension in SVE. That condensation creates a ripple slope—a tension gradient across the structured field.

What appears as curvature in a geometric sense is, in fact, the dynamic realignment of ripple patterns in SVE.

- The "falling apple" doesn't move through bent space.
- It shifts its ripple alignment along a tension slope in the field.

Curvature is not a property of a fictional fabric—it is the appearance of alignment changes across a real, structured substrate.

A Simple Analogy

Imagine watching a ball roll down a hill. To you, it looks like the ball is moving. But now, imagine that instead of the ball rolling, the hill itself is reconfiguring its slope beneath the ball, nudging it gently forward—not by force, but by gradient.

This is the UED view of gravity:

- Not curved geometry.
- But ripple slope through SVE tension fields.

You don't need a fabric to stretch. You need a structured medium to realign.

The Deeper Correction

Einstein was not wrong. He simply stopped one step short.

- He modeled the behavior of gravitational phenomena.
- But he never grounded the behavior in a physical medium.
- He used geometry to simulate a substance that was never defined.

UED fills that gap—not with metaphysics, but with a field ontology. It gives you the floor beneath curvature, the carrier behind light, and the structure behind all observed effects.

"Einstein described the bends. UED reveals the field they bend through."

The Field is Intelligent

Beyond curvature, beyond particles — the universe thinks in tension.

For centuries, science treated the cosmos like a machine: cold, calculable, and passive. Every action needed an external force. Every reaction, a measurable push. But as equations deepened and observations grew stranger, an uncomfortable question began to surface:

What if the universe isn't just reacting — but responding?

In Universal Energy Dynamics (UED), the answer is clear:

The field is intelligent.

Not in the anthropomorphic sense. Not as a conscious being with a brain or will. But in the truest physical sense: it self-regulates, harmonizes, and responds to disturbance with structured consequence.

This intelligence is embedded in Structured Vacuum Energy (SVE) itself.

Intelligence Without Thought

Let's redefine intelligence, stripped of biological baggage.

- Not memory.
- Not decision-making.
- Not abstract thinking.

In UED, intelligence means:
The inherent ability of a system to maintain harmony through dynamic self-correction.
SVE does this flawlessly.

- It balances ripple disturbances without breaking.
- It absorbs tension and redistributes energy toward realignment.
- It regulates field interactions so that no event occurs without consequence.

This is not dead space. This is living order — a substrate with laws that act as principles of intelligent symmetry.

Perception: A Ripple-Based Dialogue

In this view, perception isn't passive either. You don't just observe light, sound, or force. You align with ripple events that activate your field in structured ways.
UED proposes that:

- Perception = a ripple handshake between your field and the universe's.
- Experience = the resonance pattern formed during this alignment.
- Time = the delay or threshold it takes to achieve such alignment — what UED calls ATT (Alignment Threshold Time).

You are not seeing *reality*. You are sensing its consequence through ripple echo.

Consciousness: The Interface Itself

Where does consciousness fit?

Not outside physics. Not in mystery. But in the exact place science refused to look — the interface zone between an observer's ripple field and the SVE background.

You are not *in* the universe. You are *of* the same structure — a field fluctuation capable of sensing, aligning, and responding.

Thus, consciousness is:

The active zone of ripple interplay between structured intent and universal equilibrium.

It is not separate from the field. It is what the field becomes when localized alignment gets self-aware.

All Forces Are Field Tensions

In Newtonian physics, force is action-at-a-distance. In Relativity, it's curvature-based redirection. In quantum theory, it's exchange particles or virtual fields.

But in UED?

All forces are ripple gradients.

- Gravity = descending a ripple slope of tension.
- Electromagnetism = opposing or attracting ripple spins.
- Inertia = resistance to ripple pattern distortion.
- Even desire, fear, or memory = field tensions inside conscious ripple zones.

There is no "pulling" or "pushing"—just realignment through field consequence. Nature doesn't need cables or carriers. It uses structured imbalance.

The "God Particle" Was Never a Particle

Mainstream science searched for the "God particle"—the Higgs boson—to explain mass. But it only added another patch to the patchwork of forces.

UED reframes the search:

The true God particle is not a thing. It is the field itself—Structured Vacuum Energy—the unsplittable unity behind all division.

SVE is indivisible because it is not made of parts. It is pre-part, pre-matter, pre-force. It holds the universe the way silence holds music — not as a container, but as the condition for harmony.

When you disturb this field, you get particles. When you question this field, you get perception. When you align with this field, you become aware.

Closing Insight

We once believed in a clockwork universe. Then in a curved fabric. Now we look beneath both—and find a field that feels.

"The field doesn't think. But it doesn't have to. It listens, realigns, and answers with perfect consequence. That is intelligence—without thought."

Closing Thought

"The universe was never built on space. It was built on the intelligence of the field beneath it."

Chapter 6: What Really Moves

Ripple fields, not objects — how energy communicates through alignment

We've been lied to—beautifully, mathematically, convincingly. Since childhood, we were taught that motion means something *goes* from here to there. A car drives. A bird flies. A planet orbits. But no one asked a deeper question:

"What actually moves in motion?"

Is it the object itself? Or is it something more subtle—something underneath the object, hidden in the field it belongs to?

Welcome to the illusion.

What we call "motion" is not a thing traveling through space. It is a shift in alignment within a structured field. Nothing actually *moves*—instead, the conditions around it change, and it *appears* to have moved. Like a shadow shifting with the sun. Like a wave rolling across a still sea. The water never leaves—yet something dances across it.

In Universal Energy Dynamics (UED), we pull back the curtain. We expose the puppet strings behind what Newton called force and Einstein called geometry.

In truth, everything that moves is a ripple. And every ripple is a story of energy finding new alignment.

This chapter is not just about redefining motion. It's about liberating physics from the myth of trajectory. You'll soon see: the universe isn't a stage where actors move. It's a field where resonance shifts—and what we call objects are just anchored ripples trying to stay in tune.

Let's begin by dismantling the oldest assumption in science: That things *go* anywhere at all.

The Illusion of Object Motion

For most of human history, movement was a mystery explained by metaphor. Aristotle believed that objects moved because some force pushed them toward their "natural place"—a worldview rooted in observation but limited by imagination. Then came Newton, who reframed motion with mathematical clarity. His laws of motion introduced force, mass, and inertia into a new mechanical picture of the universe. And from that moment on, the idea that objects move through space became the unquestioned foundation of physics.

But what if that was a helpful illusion?

In classical mechanics, motion is treated as displacement—an object located at point A changes its position over time and ends up at point B. It does this by overcoming inertia via force, covering distance in space. The entire model assumes that space is an empty container and objects are billiard balls bouncing inside it.

Even in relativity, though Einstein replaced Newton's forces with geodesics in curved spacetime, the fundamental assumption remained: things move. Whether curved or flat, space is still treated as a background across which entities travel.

Now enter UED—Universal Energy Dynamics.

From the UED perspective, what we call an "object" is not an independent thing but a ripple-stabilized zone within the Structured Vacuum Energy (SVE). It is a local pattern of field tension, harmonically held in a state of dynamic equilibrium. And this equilibrium, like a tightrope walker maintaining balance, shifts not by locomotion—but by resonance realignment with the surrounding field.

Here's the key insight: Nothing ever truly moves. The object you think is "moving" is merely re-aligning its internal ripple with a new ripple condition in the field.

Analogy:

Imagine walking across a frozen lake. You feel like you're making progress from one end to the other. But now reimagine that lake as an invisible ripple network beneath your feet—every step you take doesn't shift your body, but instead realigns the ripple pattern that stabilizes your body in that location. You're not going somewhere. You're being re-expressed by the field—like a musical note changing pitch in a song.

This makes velocity not an inherent property of an object, but the rate of ripple alignment between the object and the SVE around it. It's not about "how fast you're going," but rather how quickly your field is realigning with new harmonics in the universal substrate.

Consequence:

This destroys the very foundation of traditional kinematics.

If objects don't move, then there's no "space" to move through. No absolute position to occupy. No trajectory to plot in Cartesian coordinates.

Instead, what exists is a fluid network of ripple relations, each determining how a structure is expressed in a given moment.

So when you throw a ball, you're not launching a solid object through the air. You're initiating a cascade of ripple realignments, where the ball remains itself—but its field context keeps changing in harmony with surrounding tension gradients.

In this new light, motion is not travel—it is a perceptual consequence of shifting alignment.

Next, we'll examine what's *really* moving: not objects, but ripples themselves. Let's decode Ripple Dynamics and how energy actually propagates in the absence of motion.

Ripple Dynamics – How Energy Propagates

For centuries, we imagined energy as something that traveled—waves crashing on shores, light shooting across the cosmos, sound riding the wind. From Newton's mechanical universe to Maxwell's electromagnetic fields, the story has always been: energy moves from point A to point B. It traverses space. It carries force. It displaces mass.

But what if that story is an illusion born of a false model?

In Universal Energy Dynamics (UED), we offer a bold correction: Nothing ever travels. Everything ripples.

What appears to us as motion—of objects, light, sound, or gravity—is in fact a ripple realignment within a stable, structured background field called Structured Vacuum Energy (SVE). The energy doesn't *move across* space. It reconfigures tension locally, causing the appearance of movement.

Let's unpack what this means.

Every Perceived Motion = Ripple Propagation

When you drop a pebble into a still pond, you don't say the water moved from the center to the edge. You say a ripple propagated. The water didn't travel. The disturbance did.

UED claims that the entire universe works this way.

Light doesn't travel from a distant star to your eye. What happens is a ripple tension in the SVE realigns across the field, creating a resonance pathway that your retina registers—instantly, once alignment conditions are met.

Sound isn't a wave pushing air molecules from the source to your ear. It is a sequence of ripple shifts in the medium—local adjustments in field pressure that recreate the disturbance wherever the ripple alignment finds resonance.

Gravity isn't an invisible force or curved space. It's a gradient in ripple slope—a pull not by attraction, but by natural drift into lower tension zones within the SVE.

Redefining Motion in UED Terms

Let's now formalize two key ripple dynamics:

1. Active Motion:

This occurs when intentional ripple tension shifts—like when a body accelerates, a thought emerges, or a charge moves. It represents a field-induced reconfiguration, requiring energy to disturb the local equilibrium of SVE.

Think of this as "conscious participation" in the ripple field.

2. Passive Drift:

This is when a system falls into an existing ripple gradient—like a body falling toward a planet or a leaf drifting down a river. No effort, no intent—just the system being carried along by existing tension patterns.

Falling is not being pulled—it's being resonantly redirected.

No "Travel"—Only Exchange

All fundamental phenomena—light, sound, force, even thought—are not particles or waves traveling. They are alignment consequences in the ripple field. Like the keys of a piano vibrating when a matching note is played nearby, events propagate through field coherence, not through mechanical transit.

There is no photon "flying" through space. There is field engagement between the source ripple and the observer's ripple. There is no gravity "pulling" you down. You're sliding along a ripple descent.

Summary Insight:

"In UED, energy doesn't go anywhere—it reappears everywhere the field permits."

What changes is not location. What moves is not matter. What flows is patterned alignment across a structured field that was already there.

In the next section, we'll explore how these ripple alignments don't just create perception—they define what perception is. Motion is no longer a cause—it's a symptom of field behavior.

Mass, Inertia, and Resistance to Realignment

We've spent centuries treating mass and inertia as intrinsic properties of objects—"how much stuff" something contains and "how hard" it is to move that stuff. Newton formalized this with equations, and Einstein extended the story with relativity. But no one asked:

What exactly *is* mass? And why does inertia exist at all?

In Universal Energy Dynamics (UED), we abandon the object-centric worldview. We don't ask what's inside a thing—we ask what field condition it represents.

Mass as Field Tension

In UED, mass is not material substance. It is a consequence of ripple condensation—a localized knot of field tension within the Structured Vacuum Energy (SVE).

Mass = the degree of field stress stabilized in a ripple node.

It's not "heavy." It's anchored—a stable zone in the ripple matrix that resists distortion unless acted upon by sufficient ripple gradients (e.g., gravitational or electromagnetic fields).

Think of it like a whirlpool in still water: The water isn't added—it's patterned. The whirlpool exists because of tension and curvature in the medium—not from extra water.

Inertia as Harmony Lock

So what, then, is inertia?

In UED, inertia is the resistance of a ripple node to realignment. Why? Because the node's internal ripple structure is in equilibrium with its surroundings. Disrupting it requires energy to break and reconfigure its harmonic balance with the field.

Inertia = the harmony-lock of a node within the SVE.

This transforms our understanding:

- Resistance to motion is not a result of sluggishness or inherent mass.
- It is the field's defense of equilibrium—its reluctance to

destabilize a coherent structure.
- You don't push objects—you must convince the field to reconfigure its pattern.

That's why "force" is not a primary entity in UED—it's an illusion born from ripple misalignment pressure.

Why Mass Doesn't Make You Heavy

In Newtonian terms, weight is just mass × gravity. In UED, however, heaviness is how deeply a ripple node is nested in a steep ripple slope of field tension.

- More field stress = stronger anchoring
- Heavier feel = more resistance to local realignment

This means two things:

1. Mass doesn't "pull" you down. The ripple slope underneath you releases alignment, and you slide into coherence.
2. You're not heavy because you have mass. You're heavy because your node sits in a dense ripple field (like near a planet).

The Inertia of Black Holes

UED's biggest upgrade comes in explaining black holes.

Near a black hole, the field tension is extraordinarily high. Ripple nodes become so compressed that realignment becomes nearly impossible. This is why even light appears to "stop"—not because of infinite gravity, but because:

The field's ripple slope is so steep that escape would require violating harmonic coherence.

Thus, inertia spikes. Not because there's more mass—but because the ripple gradients are too sharp to reconfigure without collapse.

Core Insight:

Mass is not stuff. Inertia is not laziness. Both are the field's way of saying: "This pattern works. Don't disturb it lightly."

In UED, reality doesn't resist change because it's slow—it resists change because it's balanced.

The next section will now explore how all movement—if it happens at all—follows ripple resonance paths, not classical trajectories.

Alignment Over Trajectory – The End of Newton's Arrow

For centuries, we have worshipped trajectories—as if motion were a divine arrow launched into the fabric of space. Newton's laws cast the universe as a cosmic game of billiards:

- Forces push
- Masses accelerate
- Arrows fly

And we believed every object had a path, a velocity, and a destination.

But Universal Energy Dynamics (UED) proposes a radical shift:

Objects don't follow paths. They follow alignments.

The arrow never travels—it realigns with the ripple field step by step. Like walking through a forest not on a trail, but by sensing where the wind flows freely. This isn't just poetry—it's physics.

Newton's Arrow: A Useful Illusion

In classical mechanics, motion is treated as the result of:

- A force initiating movement
- A continuous trajectory through absolute or relative space
- An object retaining its momentum until acted upon

This view was incredibly successful. It predicted planets, built bridges, guided rockets. But it missed the medium. It never asked:

What *enables* motion in the first place?

Newton assumed that if a force existed, motion would follow. But force was never a first cause—it was a mathematical plug for ripple misalignment.

UED Correction: No Paths, Only Realignments

Under UED, movement is not object travel through emptiness. It is a shift in the resonance pattern of ripple alignment in the Structured Vacuum Energy (SVE).

"Trajectory" is the illusion that emerges when successive alignments appear coherent.

Imagine a ripple node (what we call an "object") sitting in a stable pattern. Now imagine that node receiving a new ripple influence (e.g., intent, pressure, imbalance). The node doesn't *travel.* It re-aligns with the new ripple configuration—step by step—field by field.

Just like how a compass needle shifts direction—not because it was pushed, but because it's following the magnetic gradient.

Real-World Analogies:

- A leaf floating down a river isn't choosing a path—it's being carried by the flow, adjusting orientation to minimize resistance.
- A bird gliding doesn't fight gravity—it finds the lowest energy path through air ripples and thermals.
- Even your thoughts don't "move"—they arise from neural ripple changes, not trajectory.

UED says: Everything behaves like that. There is no velocity—only ripple cadence. There is no momentum—only persistent resonance.

Why Newton's Arrow Breaks Down

When does classical trajectory fail?

- At light speed: Where motion becomes impossible to track

linearly

- At black holes: Where gravity overwhelms all known paths
- At quantum scales: Where particles "jump" without crossing the space in between

Each of these failures isn't an anomaly. It's a signal—that the arrow never really existed. We just painted a straight line over a ripple stream and called it physics.

Ripple Insight:

Motion is not linear progress. It's a field trying to restore harmony.

This insight collapses the old duality of:

- Force vs resistance
- Motion vs rest
- Object vs space

Instead, we now understand all change as alignment drift in a field that holds everything in place.

Final Thought:

Newton gave us the arrow. Einstein bent it. UED dissolves it—and reveals the water underneath the flight path.

Nothing moves. Everything shifts. Not through space—but through resonance with an intelligent field.

Communication Through Consequence

In classical physics, information is said to "travel." A photon departs. A signal moves. A wave crosses the medium. But what if that's never what actually happens?

What if nothing is sent... only consequence is felt?

In Universal Energy Dynamics (UED), communication—be it light, sound, gravity, or thought—is not the transmission of entities, but the reconfiguration of ripple alignment between nodes that are already entangled through field conditions.

Nothing Moves — Yet Everything Responds

You strike a bell in one room, and a tuning fork in the next begins to hum. Did the sound travel? No. The tuning fork responded because it was already aligned to the same resonance.

This is how ripple communication works under UED:

No signal flies across the void. The void responds because it was prepared—charged, tuned, and consequence-ready.

We don't send signals through a medium. The Structured Vacuum Energy (SVE) doesn't carry energy like a courier. It holds every possible alignment configuration — and what we perceive as a message emerges only when a new field condition satisfies the ripple equation.

Matching Fields = Communication

Communication happens when:

- Ripple gradients match in tension,
- Field structures allow minimal resistance,
- And nodes (particles, bodies, minds) resonate in compatible frequency zones.

This isn't about a force "reaching" another particle. It's about consequence propagating because the medium itself adjusts across its structure, creating instantaneous perception.

Like birds moving in a murmuration—not because one told the other what to do, but because the field tension changed, and all participants adapted together.

Action at a Distance — Reframed

Magnetism. Gravity. Entanglement. These phenomena have long puzzled physicists because they violate our "projectile" model of reality.

- If gravity pulls the Earth instantly, what carries that pull?
- If magnets attract, where's the message?

- If particles entangle across galaxies, what travels?

The answer: Nothing travels.

These are not forces—they are ripple re-alignments in the SVE. The field didn't send a packet—it simply resolved a tension that was always present, just awaiting a trigger.

The Observer Is a Participant

Perhaps the boldest claim of UED is this:

You don't observe motion. You co-create it.

Every act of perception is a field entanglement event. You are not separate from what you observe—you are one pole of a ripple exchange.

Light doesn't "arrive" at your eye. Your visual field resonates when the alignment condition meets the ATT threshold (Alignment Threshold Time). Perception is not the passive reception of reality—it is the final ripple of a two-way handshake.

Final Insight

Communication is not the journey of a signal. It is the rippling consequence of mutual readiness.

Like a thought emerging from silence, like a feeling you can't explain, communication in UED is the art of alignment — not the motion of matter.

Stillness as Ultimate Motion

Stillness has been misunderstood.

To the classical mind, *stillness* means zero speed. A resting ball. A motionless planet. A paused pendulum. But in Universal Energy Dynamics (UED), stillness is not the absence of motion—it is the completion of motion. Not the starting point, not the stopping point, but the perfect state where ripple tensions cancel, and the field no longer needs to move.

The Greatest Illusion: Stillness as Inactivity

Physics often treats stillness as trivial—an object at rest, a reference frame, a neutral point. Yet what we call rest is simply a lack of observed displacement—not a true equilibrium of forces.
Even a rock on the ground is immersed in ripple tension:

- Gravitational fields pulling it.
- Electromagnetic fields structuring it.
- Quantum fields buzzing beneath it.

So is it really still?
UED says no.
Stillness is not the absence of energy. It is the perfect alignment of ripple gradients where net misalignment = zero.

Motion as Misalignment

Why does anything move?
Because it's misaligned with the surrounding ripple field. Motion arises from:

- A tension imbalance.
- A ripple gradient pulling toward equilibrium.
- A field condition trying to return to symmetry.

Like a marble rolling to the bottom of a bowl—not because it wants to move, but because it cannot stay imbalanced.
In UED, everything that appears to move is simply seeking rest. Not rest in the Newtonian sense of "stopping," but rest in the energetic sense of matching the surrounding field.

Cosmic Equilibrium ≠ Static Vacuum

Mainstream cosmology describes the vacuum as mostly empty and silent, with occasional particles popping in and out.
UED redefines this:
The Structured Vacuum Energy (SVE) is not silent—it is harmonically aligned.

Stillness here does not mean inactivity. It means no need to ripple. No friction. No tension. No asymmetry left to resolve.

Think of a calm lake—not because nothing touches it, but because every ripple has already resolved into resonance. That's the universal ideal.

Stillness Is the Destination

The irony is this:

All motion is a journey toward stillness.

Particles spin, planets orbit, light ripples—not to continue forever, but to seek alignment. The end of time is not destruction. It is perfect coherence.

Stillness is not a state we observe. It's a state we feel—in deep peace, in profound meditation, in awe. That is field alignment happening within us.

Final UED Insight

"Rest" is not the opposite of motion. It is motion fulfilled.

Velocity doesn't measure activity. Stillness doesn't mean stagnation. The most profound motion in the cosmos is the motion toward no motion— the ripple finally finding its resonance, the field settling into unity, the observer merging into what always was.

Closing Thought:

"The universe never moves. We do — not through space, but through alignment."

Motion is not what we think it is.

We believed in arrows, trajectories, orbits, and force vectors—believed that matter jumps across space like actors on a stage. But what if the stage never moved? What if the actors never moved either? What if movement itself is an illusion caused by changing alignment with the field beneath?

In Universal Energy Dynamics (UED), motion is not displacement—it is resonance. You don't go anywhere. The field around you reshapes, and you realign. You're not walking through the world. You're surfing through ripple states—zones of coherence shifting around your consciousness.

The rock that rests on the ground? It isn't still because gravity holds it—it's still because its ripple state matches the Earth's field tension. The photon that "travels" across galaxies? It never traveled. It simply completed an instant ripple handshake between two field alignments.

Everything we called velocity is a measure of how quickly alignment shifts, not how fast something flies. And the ultimate goal of all movement? To end the ripple tension. To arrive—not at a place, but at a state. A state of harmony. Of stillness. Of field equilibrium.

So in the deepest sense, nothing truly moves.

Only you move—not your body, not your objects, but your perception ... your field ... your resonance ... your awareness.

The universe stands still—because it is complete.

We move because we are unfinished alignments, resolving back into unity.

This chapter has shattered the kinetic illusion. But there's more.

If nothing moves, then what creates the experience of change? What stretches one moment into another, or makes some seconds feel like hours?

That mystery is not solved by clocks or rulers. It begins in perception—and emerges as something we call time.

In the next chapter, we leave the illusion of motion behind and confront the next great myth:

That time flows.

Instead, we'll see how time forms—from ripple friction, from misalignment, from unresolved resonance. Time doesn't run. It emerges—and you are its source.

Chapter 7: Time Doesn't Exist, It Emerges

Redefining Time as Alignment Threshold Time (ATT) and Perception Delay

If time were real, you'd feel it even in silence. But you don't. You only feel time when something changes. Which means—what you call time... is just your effort to keep up.

We've worshipped clocks, feared aging, and built entire sciences on the illusion of time as an independent flow—an arrow, a river, a fourth dimension. Yet, nobody has ever touched time. You've only ever felt moments stretching or collapsing inside your mind.

Einstein made it relative. UED makes it *consequential.*

In this chapter, we won't just question the fabric of time—we'll peel it back to reveal what truly moves beneath: ripples. We'll introduce a new principle—Alignment Threshold Time (ATT)—and show how your perception of change, not any ticking mechanism, creates the illusion of time.

Because the universe doesn't have a clock. It only has alignment. And that's what you've been calling "time."

The Myth of Timeflow

Time—our most trusted illusion.

For centuries, time has been portrayed as a flowing entity. In Newtonian mechanics, it was absolute—an invisible, perfectly ticking clock running in the background of all motion. It needed no justification. It simply "was."

Then came Einstein.

Relativity shattered the uniform tick of Newton's universe, revealing that time bends, stretches, and warps depending on how fast you move or how close you are to mass. No longer absolute, time became relative—but still real, still dimensional, still something.

Across both paradigms, one assumption survived:

That time *exists* as a fabric, flowing independently of mind or matter.

But what if that assumption itself is the error?

The UED Reframe

In Universal Energy Dynamics (UED), time is not an entity. It's not a river, dimension, or backdrop.

Time is a consequence. A byproduct of ripple alignment. A delay between cause and perception, determined by how quickly your field resonates with the ripple states around it.

We call this measurable ripple delay: Alignment Threshold Time (ATT).

Time emerges the moment your field begins to engage with change—not before, not after. It's not a container in which things happen. It's the friction you feel while trying to align with what's happening.

Not a Medium — A Symptom

Here's the key analogy: Just as heat is not a substance but a condition—a measurable expression of energy transfer— time is not a substance, but a perceptual symptom of field realignment.

You don't "enter" heat or "move through" it. You experience it.

Same with time. You don't "move through time." You experience ripple delay, interpreted as duration.

The Shared Flaw in Classical and Relativistic Time

Paradigm	Assumption	UED Critique
Newtonian Time	Time is absolute and universal	Mistakes measurement for substance
Einsteinian Time	Time is flexible and frame-dependent	Still assumes time is "something"
Chavan's UED Time	Time is an emergent delay	A result of ripple friction, not a fabric or flow

Ripple Insight:

"Time is not the river. You are not the boat. You are the tension in the water—and the moment you feel it shift."

In the next section, we'll explore how this ripple delay becomes measurable as ATT, and how different observers—like clocks, minds, and machines—each experience time differently based on their field structure and tension.

What Is ATT? (Alignment Threshold Time)

Time, in the Universal Energy Dynamics (UED) framework, is not an external meter—it is an internal signal.

At the center of this understanding is Alignment Threshold Time (ATT)—the minimum duration required for a ripple field to register, realign, and respond to change.

In simple terms:

ATT is the time your field takes to feel and register a shift.

Defining ATT

Every observer—whether it's a human, a machine, an electron, or an entire galaxy—exists as a ripple configuration in Structured Vacuum Energy (SVE). When change occurs in the field, no perception is instantaneous. There is always a lag—however minuscule—before the system senses, integrates, and realigns to that change.

That lag is ATT.

Formally:

ATT is the minimum ripple reconfiguration duration required for a field to perceive a change.
It is not a universal constant. It is observer-specific, state-dependent, and contextual.

Every Field Has Its Own ATT

Field Type	ATT Range	Consequence
Human brain	~80 milliseconds	Fast perception, agile awareness
Photoreceptor cell	~1-2 milliseconds	Delayed visual resolution
Quantum particle	Near-zero ATT	Seemingly "timeless" interactions
Massive planet	Extremely high ATT	Responds over cosmic time scales

Just as organisms have different reaction times, every ripple structure has its own ATT profile. This is what causes subjective time to differ between:

- The elderly and the young
- A fly and a human
- A clock in a gravity well vs. one in open space

Fast Perception = Less Time Experienced

Here lies the paradox: If your ATT is very short, you align rapidly. This means you register more events per unit of realignment, so less time feels like it has passed.
This is why "time flies" in flow states, during emergencies, or at peak awareness. Your field friction drops, ATT compresses, and the internal perception of time vanishes.

Slow Alignment = Stretched Time

On the flip side, when you're:

- Tired
- Depressed
- In a gravity well
- Or resisting change emotionally or intellectually

Your ATT increases.

You take longer to register change, and as a result, time feels slow. A single minute in boredom can feel longer than an hour of inspired creativity.

This has nothing to do with actual events and everything to do with your internal ripple tension.

Why Clocks Can't Measure ATT

A mechanical clock ticks in uniform intervals based on atomic transitions or mechanical oscillations. But it doesn't "know" what change feels like. It doesn't perceive—it just cycles.

ATT is a perceptual measurement, not a mechanical one.

You don't measure ATT from outside—you experience it from within.

In UED terms:

Time is not what your clock shows. Time is what your ripple takes to adapt.

Core Insight:

"Your ATT is your personal time signature. You are not in time—you are time in the act of aligning."

In the next section, we'll explore how this internal alignment becomes visible as ripple delay in observation, and how this delay—not velocity—is what distorts perception across frames.

The Role of Perception in Creating Time

Time is not a *thing* you sense— You never "see" time, "touch" time, or "move through" time.

What you actually perceive is change—and then your mind labels it with duration.

In the UED framework, time is not an objective flow or cosmic backdrop.
Time emerges as a side effect of perception—a ripple artifact born from your field's effort to register reality.

How Time Emerges in UED

In Universal Energy Dynamics, your perceived time forms when three events converge:

1. Observer Intent (your ripple state)
2. External Field Disturbance (change in your environment)
3. Ripple Realignment (the field's effort to restore coherence)

If alignment is instant, time doesn't register—there is only presence. If alignment is delayed, your field struggles to catch up—this struggle creates the sensation of time.

Time = ripple resistance felt as delay.

The Three Steps That Generate "Time"

Phase	What Happens	Outcome
1. Perception Initiated	Your attention/intention engages with a change in the field.	Start of realignment
2. Resistance Encountered	Your internal ripple clashes with the external shift.	Effort required for alignment
3. Experience Registered	Alignment succeeds, and memory locks it in.	Change is "felt" → Time is born

Thus, without perception, there is no time. If a tree falls in the forest and no ripple field responds to it—no time passes.

Emotion Modulates ATT — And Shapes Time

Your emotional state isn't just psychological—it modulates your ripple tension. That means it directly affects ATT and therefore how you experience time.

Examples:

- Fear or pain → High internal resistance → Slower alignment

→ Time feels stretched
    - e.g., a car accident that seemed to last forever in your mind.
- Joy or flow state → Perfect resonance → Rapid alignment → Time feels compressed
    - e.g., spending hours doing something you love but feeling like minutes passed.

This is not imagination— It's your ripple field reporting back from resonance or turbulence.

Memory as Time's Evidence

Time becomes real only when change is stored. If the field realigns without friction, no memory is needed—you were simply present.

But when resistance is high, the alignment leaves a scar—a ripple echo we call memory. And that's when you say:

"That moment felt so long."

Because it was—not in clock units, but in ripple labor.

Ripple Insight:

"Time is not the length of an event. It's the effort your field makes to digest that event."

So, your time is not the world's time. Each observer lives in a different ripple clock, defined by their own alignment threshold and field reactivity.

Core Realization:

You don't move through time.

You move with change—and time forms in your wake.

Clock Time vs. Field Time

"A clock may tell you the hour—but only your field knows the moment."

We've spent centuries building tools to measure time—but few ever questioned what exactly they were measuring.

A clock doesn't know you. It doesn't care whether you're in love, in pain, or in deep meditation.

It just ticks.

What Clocks Really Measure

Clocks are machines of mechanical repetition. They are external recorders of periodic motion—pendulums, quartz vibrations, atomic decays. Each beat is assumed to represent a fixed unit of time.

But in UED, this is temporal projection—not reality.

- A second on Earth ≠ a second near a black hole.
- A minute in boredom ≠ a minute in ecstasy.

And yet, the clock insists they're the same.

This is the great conflation:

Mistaking mechanical uniformity for lived rhythm.

Case Study: GPS and Time Correction

Take GPS satellites. Because they orbit Earth in weaker gravitational zones, they experience different ripple alignments than surface-based observers.

General Relativity accounts for this by applying time dilation corrections.

But what's really happening?

It's not that "time slows down" up there. It's that field alignment rates differ based on ripple tension zones.

- The satellite's internal rhythm is calibrated to a different field configuration.
- GR patches this discrepancy using mathematics—but never explains the medium that causes it.

UED Insight: The True Clock Is Your Field

In Universal Energy Dynamics:

- Time is not measured—it is felt.
- Your field time is defined by how quickly your awareness

can adapt to ripple change.

This means that two people in the same room may be experiencing entirely different durations, depending on their internal ripple alignment.

- The insomniac's hour feels eternal.
- The child's summer feels like a blink.

Same clock. Different fields. Different timelines.

ATT Over "T" — A New Foundation

Physics today relies heavily on fixed time parameters (T, dt, etc.) to describe motion, force, entropy, and causality.
But UED proposes a foundational shift:
Replace "T" with "ATT" — Alignment Threshold Time.
This reframes every equation not as an event in time, but as an event that emerges when ripple alignment completes.
Implications:

- Speed becomes ripple responsiveness, not distance per unit time.
- Entropy becomes ripple diffusion, not time's arrow.
- Causality becomes alignment sequence, not timestamp comparison.

The Risk of Using "t" as a Crutch

Time constants were always a convenience, not a reality.
They helped us model outcomes. But they:

- Locked perception into equations
- Excluded observer rhythm
- Flattened experience into uniform slices

It's like mapping music using a metronome but ignoring harmony, rhythm, and resonance.

Field Time = Navigating Reality by Feel

Your life isn't lived in clock time. You don't make decisions in milliseconds or Planck units.

You feel what's changing. You sense delays, resistances, accelerations.

That's Field Time — your built-in ripple clock. Not synchronized with any machine—but always synchronized with your state of being.

Clocks record time. But only consciousness creates it.

Time Dilation — A Ripple Perspective

"Time doesn't stretch—it resists. The field doesn't slow you down. It asks more from you."

Traditional Interpretation: The Stretch of Spacetime

In Einstein's General Relativity (GR), time dilation occurs when:

- An object moves very fast (Special Relativity), or
- It exists near a massive body like a planet or black hole (General Relativity).

In both cases, clocks appear to tick slower relative to other frames of reference. The common interpretation is:

"Spacetime curves and stretches, altering the flow of time."

But what does "stretching" time even mean?

GR offers mathematical adjustments but never defines the mechanism. There is no explanation for *why* clocks slow—just that they must, to match observations.

Chavan's UED Reframe: It's Not Time Stretching—It's Field Straining

Universal Energy Dynamics (UED) doesn't see time as a fabric or a flowing dimension. Instead, it offers a ripple-based reinterpretation:

What you're calling *dilation* is not time slowing— It is the field around you becoming harder to realign.

Let's break this down.

The Mechanics of Perception: ATT and Friction

In UED:

- Time emerges when ripple alignment completes.
- The ATT (Alignment Threshold Time) is how long it takes for the observer's field to synchronize with an external change.
- In areas of high ripple tension (e.g., near a black hole), this alignment requires more effort.

Think of ripple tension like friction:

- In low-density fields: Ripples reconfigure easily → short ATT → quick perception → "fast time."
- In high-density fields: Ripples struggle to reconfigure → long ATT → delayed perception → "slow time."

But here's the key insight:

The amount of time hasn't changed. Your ability to register change has.

High Mass = Steeper Ripple Slope

Massive objects don't bend time. They alter the ripple structure of the vacuum field—creating steeper ripple gradients.

This causes:

- Higher field density
- Increased ripple friction
- Slower ripple feedback
- Increased ATT

The result?

- Light appears to redshift.
- Clocks seem to slow.
- Time feels like it's "dilated."

But it's not dilation—it's alignment difficulty.

Key UED Insight: Perception vs. Duration

Let's reframe:

Classical View (GR)	UED View
Time slows near mass	ATT increases near mass
Time stretches	Ripple reconfiguration is harder
Clocks tick slower	Field response is delayed

This changes everything.

- You're not experiencing *less* time.
- You're experiencing a *harder time aligning.*

It's like trying to run underwater—you're moving, but resistance makes it feel slow.

Analogy: Streaming in Slow Internet

Imagine watching a video on slow internet:

- The server didn't slow down.
- The content didn't change.
- But buffering delays the experience.

That's time dilation in UED. The ripple content exists—the field is simply buffering your perception of it due to high local tension.

Field Tension Artifact, Not Dimensional Distortion

This is a profound shift:

- No spacetime needs to bend.

- No metaphysical "clock of the cosmos" needs to stretch.
- Just ripple mechanics, working against a tense field.

Time dilation is a symptom, not a fundamental truth.
It's what happens when the field resists realignment.

Final Reflection

"In truth, time doesn't slow. Your awareness lags behind the ripple. The delay you feel... is the field asking you to catch up."

Memory, Expectation, and the Loop of Time

"Time feels real because you remember something, expect something, and forget that both are fabrications."

Why Time Feels Like It's Flowing

Humans don't perceive time directly. We perceive change, then we string together:

- What we remember (the past)
- What we expect (the future)

The result?
A loop of perceived continuity that feels like "time is moving forward."
But this continuity isn't fundamental. It's assembled—like a movie created from still frames.

Memory: The Echo of Ripple Alignment

In UED, memory is not stored in a brain like files in a cabinet. Instead, it is the resonant echo of prior ripple alignments within the field of the observer.

- Every experience is a ripple configuration.
- Memory = a stored alignment pattern in your internal field.
- Recalling something is replaying that resonance.

This explains why memory is subjective, emotional, and sometimes unreliable. It's not a recording—it's a field replay.

Expectation: The Future You're Trying to Align

Expectation isn't just prediction. It is a ripple projection—your field attempting to align with a future possibility.

- When you expect something, your internal field pre-configures itself.
- This influences perception, reaction, and emotion—even before the event occurs.

In UED, expectation is energy spent forward, a kind of ripple scaffolding built toward a future resonance.

Time = The Tension Between Memory and Expectation

The moment you exist in is caught between:

- The echo of what was (memory)
- The projection of what might be (expectation)

This creates temporal friction. That friction is what you *feel* as time.

"You are always resisting the past and reaching for the future—creating the illusion of flow."

It's not time that flows. It's your field, sliding between ripple echoes and future alignments.

Analogy: The Mind as a Film Editor

Imagine a film editor working with raw footage (memory) and a storyboard (expectation). Their job is to cut, align, and splice moments together into a coherent narrative.

- That narrative is what we experience as "time."
- But the editor isn't time—it's you.

Your awareness loops between past ripples and projected ones, keeping the reel moving.

The Feedback Illusion

Every new alignment you experience:

1. Updates memory
2. Reinforces expectation
3. Strengthens the illusion of continuity

It's a self-looping system.

- The past feels fixed because the memory pattern is strong.
- The future feels real because the expectation ripple has momentum.

But all of it is feedback in a ripple field.

Time Loops and Distortion

When either memory or expectation dominates, the loop gets distorted:

- Trauma = overcharged past ripple → memory loop traps you
- Anxiety = hyperactive future ripple → expectation loop overwhelms
- Flow state = no past or future interference → field aligns in real-time → time disappears

Thus, the quality of your field determines your experience of time.

UED Formula (Conceptual)

Let's condense it:

- Memory = Past ripple alignment echo
- Expectation = Projected future alignment attempt
- Time experienced = Friction between both

When friction is high, time feels heavy. When friction is low, time vanishes.

Final Reflection

"Time is a side effect of trying to hold the past and predict the future while forgetting to just align."

When Time Stops

"Time doesn't stop—it stops *forming.* Because you stopped feeding it."

Can Time Truly Stop?

We've all felt it:

- In deep meditation, there is no sense of duration.
- In a flow state, hours pass like seconds.
- In a moment of trauma, seconds stretch like hours.

These are not anomalies—they are *clues*. They reveal that time isn't fundamental—it's field-dependent.

And sometimes, your field stops generating it.

ATT and the Spectrum of Time Perception

UED defines time as an emergent effect—specifically, the delay between field realignment and perception, or:

ATT — *Alignment Threshold Time*

Two extremes define the edge of time:

1. ATT → Zero (Time Disappears)

- Occurs in deep flow, meditation, or transcendence.
- Your internal field aligns instantly with the external ripple.
- There is no resistance, no delay, no memory, no expectation.

You don't feel time because your perception is perfectly synchronized with the field. The mind isn't waiting for feedback—it's dancing *with it.*

This is the true "now." Not a moment in time—but a resonance.

2. ATT → Infinity (Time Freezes)

- Happens in trauma, shock, or hypervigilance.
- Your field gets overwhelmed—ripples loop without resolution.
- The observer is flooded with feedback but cannot align.

Result:

- Time seems to *slow* or *freeze.*
- Every second is processed in excess.
- You feel stuck, paralyzed—not because time changed, but because your perception can't realign.

The mind isn't frozen in time. It's stuck in a ripple loop that won't collapse.

What Does It Mean to "Stop Feeding Time"?

Time forms only when you:

- React to change.
- Struggle to realign.
- Record the delay.

This means your response *generates* the feeling of time.

In meditation, or deep creative absorption:

- You stop reacting.
- You stop anticipating.
- You stop remembering.

You stop creating internal ripple delays—so time, as you know it, ceases to emerge.

Trauma vs. Transcendence: Two Sides of the Same Ripple

Interestingly, both trauma and transcendence bend time—but in opposite ways:

State	ATT Value	Experience	Field Condition
Trauma	Approaching ∞	Time freezes	Ripple overload; misalignment
Meditation	Approaching 0	Timeless presence	Perfect ripple alignment
Flow	Minimal ATT	Time flies	Low-friction resonance

In each case, *you didn't escape time—you altered how much ripple delay your field allowed.*

Consequence: Time Requires You

Time isn't just about what's happening—it's about how you process it.

- If you're aligned, no time.
- If you're overwhelmed, infinite time.
- If you're recording, tracking, reacting—you're building time.

Time is not universal—it's your participation in ripple response.

Final Insight

"You don't need to stop time. You just need to stop feeding it ripple resistance."

The most timeless beings aren't outside the world. They're inside it—so deeply aligned that time doesn't get a foothold.

Closing Reflection

"Time is not a river you float through. It's the wake you leave behind in a sea of ripples."

For centuries, we've imagined time as something external. A river we are thrown into at birth. A line we march across. A ticking clock indifferent to our lives.

But under Universal Energy Dynamics (UED), this metaphor collapses.

Time is not flowing past you. You are creating the illusion of flow—through your own ripple trail.

The Observer's Wake

When you move through the field, you do not "travel in time." You interact, realign, and respond. Each ripple of perception you generate—through thought, emotion, decision, awareness—leaves a trace in the SVE (Structured Vacuum Energy).

These traces echo back at you as memory. They build expectation. They give the illusion that you are moving forward on a timeline.

But really, you're just looking behind—watching the pattern you already made.

You Don't Float Through Time—You Ripple It

Imagine the universe as a still sea. Your intent, focus, and attention disturb the surface. That disturbance is what you call experience. And when the ripples reflect back at you—you call that time.

You aren't a passenger. You are the source.

Why This Matters

The implications are vast:

- Time doesn't act on you.
- You act, and time forms in response.

This empowers you.

It means healing can reshape your past (by diffusing ripple loops). It means presence can dissolve future anxiety (by ceasing unnecessary ripple projections). It means stillness is not idleness—it's mastery of the field.

The New Time Equation

Forget "t" as a constant. Forget clocks, calendars, cosmic timelines.

In UED, time is:

Time = Consequence of Perception Delay (ATT) = *Friction* + *Feedback* + *Field Misalignment*

Which means...

No misalignment → No friction → No time.

Final Thought

Time is not something you live inside. It's something that lives inside your ripple trail. And once you stop feeding it... you don't grow old. You grow clear.

Chapter 8: Light Isn't Traveling

A ripple that aligns to illuminate — not a photon in flight

Let's begin by unsettling the most cherished certainty in all of physics: What if light doesn't travel at all? Not as a particle, not as a wave—not even as a probability cloud surfing spacetime. What if every flash, glint, and ray you've ever seen is not an object in motion but an event in alignment?

We've imagined light as a courier—departing from a distant star, weaving through the cosmic void, and arriving dutifully in our telescope lens. We've charted its speed, revered its constancy, and used it to measure the universe itself. And yet—no one has ever seen light in flight.

What we call "light" is not something traveling across space. It is what happens when two ripple fields meet in resonance across the Structured Vacuum—when the source and observer momentarily agree to illuminate. No particle moves. No wave crosses the void. Only alignment occurs—and perception ignites.

This chapter unravels the illusion of photon motion. It proposes a radical reframe: Light isn't something that arrives. It's something that happens—at the point of mutual field readiness.

Let go of the beam. Let go of the flight. Because in truth, you never saw light move. You only saw yourself aligned—perfectly—with a ripple already waiting.

The Legacy of the Photon

Before we challenge the idea that light travels, we must understand how that idea became sacred in the first place.

From Corpuscles to Waves to Photons

The story of light has always been a story of reinvention. Newton once imagined light as tiny corpuscles—discrete particles emitted from luminous bodies. It was a mechanical model, elegant for its time, fitting neatly into the classical framework of cause-and-effect physics.

Then came Maxwell, who swept away the particle image with the majesty of electromagnetic waves. Light, he argued, was nothing more than a self-sustaining oscillation of electric and magnetic fields, propagating through space at a constant speed—c. The world rejoiced. Light was now not a thing, but a pattern.

But this harmony was shattered in the early 20th century when Einstein explained the photoelectric effect by suggesting light came in packets—quanta. These were later dubbed photons: discrete energy bundles that acted like particles but often behaved like waves. Physics was split—not just conceptually, but ontologically.

Thus, the modern view emerged: light is both a particle and a wave, depending on how you look at it. But no one asked the deeper question: What exactly is moving?

Why Light Became a Traveler

The idea of light as a traveler wasn't born from observation—it was born from interpretation. Because we receive light from distant stars, we assume it must have traveled. Because it affects detectors, we assume it must have carried energy. Because we measure time delays between events, we assume light moves through space at a finite speed. And because no other model fit the dominant geometry-centric view of physics, the traveler narrative stuck.

But these assumptions were layered onto a model—not derived from any direct interaction with a moving photon. The truth is: no one has ever seen a photon move. We only observe emission and detection—two ends of an alignment.

The Duality Illusion

The wave-particle duality that dominates modern quantum theory is often presented as a profound mystery. How can something be a spread-out wave and a pinpoint particle at once? How can light interfere with itself as a wave, but strike a surface as a particle?

The answer may be simpler than expected—and more radical. Perhaps it isn't either. Perhaps duality isn't the nature of light—it's the nature of our perception. A perceptual artifact—born from observing incomplete interactions and assuming continuity where none exists.

UED Insight:

Light is not a contradiction—it is a ripple resonance. The duality vanishes when we stop imagining light moving from point A to B and instead recognize it as a moment of field agreement between A and B. The wave? That's how the field prepares for resonance. The particle? That's how we register it.

What We Actually Observe

Let's pause and reflect on a deceptively simple question: Have you ever seen light move?

The honest answer is—no one has.

We Only See the Start and the End

Every observation of light in scientific history has occurred at one of two points:

- The moment it is emitted (e.g., a bulb turning on, an electron de-exciting),
- Or the moment it is detected (e.g., hitting a retina, a photographic plate, or a telescope lens).

The space in between? That's never observed. It's inferred.

We assume a photon was "in transit." But what if this assumption is the problem? What if nothing is traveling—and something else entirely is happening?

Detection is an Alignment Event

Whether it's sunlight on your skin, a laser striking a mirror, or a cosmic ray reaching a satellite, the interaction is always localized. What we call "light" is not a continuous journey but a punctual handshake between two ripple fields: the emitter and the observer.

These events are alignment-based. They occur only when the field tension, resonance, and ripple thresholds between the source and receiver match. It's not a photon "arriving"—it's two regions of Structured Vacuum Energy (SVE) becoming momentarily synchronized.

The Myth of the Traveling Photon

We teach students that photons travel from the Sun to Earth in ~8 minutes. But what actually happens?

- Do we track the photon?
- Do we watch it flying through space?

No. We record a delay and assign that delay to "photon travel time." But this is just a causal gap—not a physical proof of movement.

In every lab, telescope, or optical experiment, photons are not seen in flight. They are seen at endpoints—at the intersection where ripple meets resonance.

Ripple View: Instantaneous, Not Transported

UED proposes a fundamental reframe: Light is not a thing moving across a void. It is a ripple alignment across a structured field. Like two tuning forks vibrating together across a shared medium.

- The Sun's emission field and your eye's detection field align across a ripple slope in SVE.
- When the alignment threshold (ATT) is reached, illumination happens.

The delay we observe is not because light was in transit—it's because the ripple alignment took time to lock in, based on field readiness and distance-related resistance.

UED Insight:

"Light isn't something that travels. It's something that happens—when the universe agrees to let you see."

This changes everything: It removes the need for photons as messengers, space as a carrier, and speed as a limiter. It opens the door to understanding perception itself as a ripple phenomenon.

Ripple Alignment, Not Propagation

In conventional physics, we are taught that light propagates—it moves through space either as a particle (photon) or as a wave (electromagnetic oscillation). This idea underpins everything from communication networks to our understanding of the universe's early moments.

But what if this foundation is misread? What if the "propagation" of light is not a journey, but an event—a field-level handshake between points already immersed in a dynamic, structured medium?

UED's Radical Reframe: Light Doesn't Travel—It Aligns

Under Universal Energy Dynamics (UED), light is not a substance that moves—it is a ripple field event. This event occurs when Structured Vacuum Energy (SVE), the universal background of all existence, facilitates an alignment between emitter and observer.

- There's no need for a particle to fly through space.
- There's no wavefront sweeping forward through emptiness.

Instead, two fields—each already embedded in the SVE—reach a resonance state across the ripple substrate. Once Alignment Threshold Time (ATT) is achieved, perception occurs.

Analogy: Sync, Not Transfer

Imagine two people standing at opposite ends of a perfectly still pond.

- They each begin clapping in perfect rhythmic sync.
- From a distance, the sound appears simultaneous—no delay, no visible cause and effect.
- But nothing has "moved" across the pond.

Now replace sound with ripple resonance. The still pond becomes the SVE field. The clapping? Ripple alignment—a real-time, mutual field readiness between source and observer.
In this model:

- The Sun doesn't "send" light to Earth.
- The Sun's ripple field and Earth's observational field align across SVE.
- Illumination is not transferred—it's mutually triggered.

Ripple Fields Carry Potential, Not Packages
In SVE, there are no moving photons—only potential ripple states waiting for resonance. Each point in the vacuum holds latent capacity to resonate, to ripple, to become illuminated—not through motion, but through tuned field alignment.
This is why:

- No one sees light in transit.
- Photons are always "born" at detection.
- Emission and absorption happen together across a shared ripple gradient.

ATT: The Window of Perceived Instantaneity
Alignment Threshold Time (ATT) defines how quickly an observer's field can register ripple resonance. Within this window:

- The ripple appears instantaneous.
- No motion is required or even meaningful.
- What we interpret as "speed of light" is really the delay in field readiness—not a particle's velocity.

So when we measure c = 299,792,458 m/s, we're not clocking a traveler—we're quantifying the minimum ATT threshold required for alignment between emitter and observer in a low-resistance field.

Ripple Insight:

"Light doesn't cross the distance. The distance dissolves when ripples align."

This single shift—from propagation to alignment—reshapes everything we think we know about optics, causality, time, and even reality.

Rethinking Speed of Light (c)

For over a century, the speed of light—denoted as "c"—has been enshrined as the ultimate speed limit of the universe. Nothing, we are told, can move faster. This belief forms the backbone of Einstein's relativity, quantum field theories, and nearly every pillar of modern physics.

But what if this "speed" isn't speed at all?

What if "c" is not how fast something moves, but how long it takes your field to perceive a ripple alignment?

From Speed to Perception Delay

Under Universal Energy Dynamics (UED), the speed of light is not the velocity of a particle or wave. It's a temporal threshold—a delay window in ripple field realignment that makes perception possible.

Let's unpack this:

- When we say "light takes 8 minutes to reach Earth from the Sun," we assume it's traveling through space like a bullet.
- But we've never observed this journey—we only detect the

light once it hits our instruments.

- In UED, that delay is not transit time—it is the Alignment Threshold Time (ATT) required for the observer's field to resonate with the Sun's ripple field across the SVE.

So, c is not how fast light goes— It's how long alignment takes under current field conditions.

What Appears as Speed Is Actually Alignment Latency

This radically shifts the concept of motion:

- From emitter to observer, no particle moves.
- What we perceive as "arrival" is really the completion of ripple realignment across a structured vacuum field.
- The ripple itself is already present in the SVE. It does not "travel"—it activates.

This also explains why:

- Light always appears to move at the same speed, regardless of observer motion (as confirmed by the Michelson-Morley experiment).
- The so-called "invariance" of light's speed is really the invariance of ripple resonance delay across aligned fields—not some magical cosmic law of speed.

Implications for Physics

By redefining c as a field-dependent threshold, we open the door to:

1. Field-centric modeling of perception, causality, and resonance.
2. An updated view of time dilation, redshift, and even quantum entanglement—not as bizarre spacetime deformations or faster-than-light paradoxes, but as ripple

delay dynamics.

3. A rejection of "light travel" in favor of light emergence—where perception is a shared event between source and observer, not a signal crossing distance.

Analogy: Light as a Ripple Match, Not a Messenger

Imagine two tuning forks—one with the source, one with the observer. When the ripple tension and frequency align through the medium (SVE), the second fork resonates.

It doesn't need a particle to fly from one to the other. It just needs ripple harmony across the field. That delay—however small—is your measured c.

Ripple Insight:

"Light doesn't race across the void. The void realigns—then illumination appears."

Light Bending — A Ripple Phenomenon

One of the most celebrated confirmations of Einstein's General Relativity came in 1919, during a solar eclipse. Arthur Eddington's team observed that starlight appeared to bend as it passed near the Sun—seemingly validating the idea that massive bodies curve spacetime, and light follows that curvature.

But what if this was a misinterpretation of the mechanism—right result, wrong reason?

Revisiting the Phenomenon with Ripple Eyes

Under Universal Energy Dynamics (UED), the Sun does not bend "space" like a trampoline distorting under a bowling ball. There is no invisible fabric. Instead, the Structured Vacuum Energy (SVE) around a massive object like the Sun is denser, forming a ripple slope—a zone of increased field tension.

In this model:

- Light doesn't "curve" in the traditional sense.

- Ripple paths realign in response to the gradient of SVE field tension.
- This alters the apparent direction from which the ripple (light) is perceived to come.

Thus, gravitational lensing is not light obeying geometry—it's a field effect, and the observer's perception is being redirected by ripple distortion.

Not a Force — A Preference

Unlike classical physics, where light bends due to forces, or GR where curvature acts as pseudo-force, in UED:

- Light isn't being forced to bend.
- It simply follows the path of least ripple resistance.
- The ripple field near a massive body naturally redirects alignment trajectories, like wind flowing around a mountain—not because it's pushed, but because the terrain shapes the flow.

From this view, light doesn't resist—it harmonizes. It follows the shortest ripple reconfiguration path, which is not always straight in a field with uneven SVE tension.

Perception: The Final Curve

What we call "light bending" is not the light curving—but our own alignment path shifting. The observer's field locks onto the distorted ripple field and reconstructs the source accordingly.

This is why:

- The starlight seems to come from a different place.
- But in reality, what changed was the alignment medium—not the location of the star.

So, the image you see is a consequence of field distortion, not a literal relocation of photons.

Analogy: Water Refraction Reimagined

Imagine watching a straw dipped in a glass of water. The straw appears bent. But it isn't the straw that changed—it's the medium that altered the path of light reaching your eyes.

Similarly, in UED:

- The "bend" is an alignment shift due to ripple tension gradient.
- Your perception reconfigures based on that shift.
- The illusion of curve is born in your observation, not in space.

Ripple Insight:

"Light doesn't bend. It aligns—through tension, not trajectory."

Quantum Jumps and Instantaneous Light

In classical physics, light travels. In quantum physics, it jumps.

Quantum Electrodynamics (QED), the most precise theory of electromagnetic interaction to date, has long hinted at a stranger reality beneath the classical illusion of motion. It describes how photons are emitted and absorbed by charged particles—but crucially, it says nothing is observed between those two points. No one has ever witnessed a photon "in flight."

What Quantum Jumps Tell Us

At the quantum scale:

- Electrons jump between energy levels in atoms.
- When they drop to a lower level, a photon is emitted.
- Another atom may then absorb that photon.

This entire interaction appears to occur instantaneously—and without a visible trajectory. Despite popular descriptions, there is no path, no mid-air photon cruise. It's as if the information transfers without movement.

Even Richard Feynman, who developed the path integral formulation of QED, admitted that every possible path must be considered mathematically—not because light takes them, but because we can't say which one (if any) is real. It's not that photons are traveling weird paths—it's that they may not be traveling at all.

UED's Reframe: No Photon in Flight

Universal Energy Dynamics (UED) offers a cleaner ontology for this quantum weirdness:

- There is no particle moving from A to B.
- What exists is a ripple field—a potential alignment structure in the Structured Vacuum Energy (SVE).
- When two field nodes (emitter and absorber) are in harmonic alignment, a ripple activation occurs.

The appearance of a photon "being emitted" is just the ripple node reconfiguring its internal field. The "absorption" is the other node syncing to that ripple resonance.

Light doesn't travel. It synchronizes.

In this view, nothing moves between the two points. The entire event is a consequence of shared ripple readiness in the SVE field.

Entanglement: The Proof in the Paradox

Nowhere is this clearer than in quantum entanglement:

- Two particles interact and become correlated.
- When one is measured, the other instantly reflects the change—no matter the distance.

Einstein called it "spooky action at a distance." But under UED, it's simply ripple co-alignment across the field. Distance becomes irrelevant when the ripple tension between the two nodes is already entangled in the SVE substrate.

Entanglement doesn't require information to move. It requires fields to realign.

Implications for Light:

- What QED and entanglement hint at, UED formalizes: Events aren't connected by travel—they're born of alignment.
- Light is not a flying photon—it is an activation event between two points with matched ripple potential.
- The speed of light ("c") is not a velocity—it's the alignment threshold time (ATT) limit for human-scale perception.

Ripple Insight:

"The quantum jump is not a jump at all. It's the sudden visibility of what was already connected."

Why We Needed the Illusion of Light Travel

For over a century, light has been described as a traveler—a photon darting through the vacuum, delivering energy from point A to point B. This imagery is not just rooted in physics textbooks, but embedded in the foundations of how humans perceive cause and effect.

But why did we accept this idea so easily? Why did we need the illusion of motion?

Let's decode this dependency—not to dismiss its utility, but to reveal the deeper field dynamics we've overlooked.

The Utility of the Illusion

From a practical standpoint, modeling light as a traveling entity was profoundly useful:

- Telescope calculations relied on light's finite speed.

- Optical lenses were designed assuming rays bending and traveling.
- Relativistic physics incorporated light's constant speed into GPS systems and satellite synchronization.
- Everyday metaphors—like "seeing the past" by observing stars—depended on the notion that light moves, carrying snapshots of history.

This conceptual convenience gave us control. It allowed predictions, blueprints, equations. And it worked—so we embraced it fully.

The Deeper Psychological Bias

But beyond practicality lies cognitive framing. Human perception is sequential:

- We see A happens, then B happens.
- We assume something must have moved from A to B.
- Our brains expect motion to bridge cause and effect.

Thus, when light from a bulb strikes your retina, it feels intuitive to imagine a particle that traveled the gap—carrying the "visual data" from source to sensor.

In truth, this may be less about science, and more about the way human minds resolve delay and consequence.

What This Illusion Cost Us

Yet in our reliance on this visual metaphor, we may have missed a subtler, more elegant reality:

- Light doesn't "bring" the world to you.
- It doesn't need to travel across empty space to deliver meaning.
- What you perceive is not a delivery—it's an activation of pre-existing potential in your field of awareness.

By imagining photons in flight, we externalized perception—as if light were an external courier. But ripple dynamics say: light is the consequence of your field coming into resonance with another.

There is no transit. There is only field alignment.

UED Insight:

"We never needed light to travel. We needed it to appear—to match, to ripple, to harmonize. What you see is not what arrived—it's what aligned."

Closing Implication:

The illusion of light travel served us well. It built physics, optics, and GPS satellites. But illusions, when mistaken for truth, become barriers to deeper understanding.

Ripple-based light requires no messenger. It requires mutual readiness, like two strings vibrating in perfect sympathy. The Structured Vacuum Energy (SVE) carries the potential for all illumination. You only see what you're aligned to receive.

Closing Reflection

"Light is not a messenger from afar. It's your ripple meeting the universe's—at just the right resonance."

For centuries, we imagined light as something *sent*—a photon flung across the cosmic void to deliver the news of the universe. A radiant courier. A traveler of infinite patience.

But what if this messenger never actually moved?

What if all it ever did was *resonate*—not across distance, but across readiness?

In the Universal Energy Dynamics (UED) framework, light is not a particle or a wave on a mission. It is a moment of alignment. A ripple event that occurs when two fields—source and observer—are ready to lock into resonance within the Structured Vacuum Energy (SVE).

Nothing flies. Nothing traverses.

Instead, something activates.

You don't *see* because something reached you. You see because you and the universe touched—through ripples, through field coherence, through mutual readiness.

This shift is not just poetic—it is consequential.

It changes how we interpret causality: Not as linear, but as conditional.

It redefines observation: Not as passive reception, but as active co-participation.

It transforms the act of seeing: Not as detecting signals, but as being in tune.

In a world where light doesn't travel, you don't wait for photons—you tune your field. You refine your resonance. You meet reality through alignment.

This is not just a new physics. It's a new ontology. A new way of understanding being.

Chapter 9: Gravity Without Geometry

Why things fall in ripple gradients — not because spacetime curves

What if gravity isn't a force at all? Not a pull. Not a curve. Not even an interaction.

Imagine dropping an apple—not because some invisible hand tugged it downward, and not because spacetime decided to dimple under its feet, but because the field beneath it *whispered an invitation.*

For over a century, we've pictured gravity as the elegant warping of a cosmic trampoline, objects rolling toward dents in the fabric. Beautiful. Intuitive. But perhaps... mistaken.

What if Einstein's curved stage was never the final truth—but only a clever rendering of a deeper ripple slope hidden beneath the script?

This chapter dismantles the illusion of geometry and reveals gravity as a passive alignment to field tension—a ripple gradient through Structured Vacuum Energy (SVE), not a force imposed or a shape bent.

Welcome to a world where things don't fall. They align.

Revisiting the Core Assumption

For millennia, gravity has been the most familiar yet least understood phenomenon in physics. From the ancient pull of falling apples to the majestic orbits of planets, we've always needed a story to explain why things fall—and stay together.

The Classical Narrative: Gravity as Force (Newton)

In Newton's framework, gravity was a force—real, measurable, and instantaneous. Every object with mass attracted every other, acting at a distance with a strength defined by a beautiful, simple equation. It worked. It predicted planetary motions. It helped launch rockets. But it didn't explain *how* the force acted. What medium? What mechanism? That question remained unasked—because the math was good enough.

The Relativistic Upgrade: Gravity as Geometry (Einstein)

Then came Einstein. In 1915, with General Relativity (GR), he shifted the paradigm. Instead of an invisible force acting across space, Einstein proposed that space *itself* bends and curves in the presence of mass and energy. Matter tells spacetime how to curve; curved spacetime tells matter how to move.

This geometric interpretation gave gravity a new elegance. Free-falling objects weren't being pulled—they were simply following the straightest possible paths (geodesics) through curved space. It was revolutionary. It explained Mercury's orbit. It predicted black holes. It matched experiments.

But it came with a silent assumption: That space—and time—are *real physical fabrics* that can warp, bend, and stretch.

Why Curvature Was Necessary in GR

Einstein needed curved spacetime because there was no other known mechanism to explain why an object in free fall would appear to accelerate without any force acting upon it. In the classic elevator thought experiment, a person falling inside an elevator feels weightless. GR explained this with curvature—gravity was no longer a force; it was the shape of the path.

Curvature replaced causality. But only by replacing substance with abstraction.

The UED Challenge: Was It a Ripple Slope All Along?

Universal Energy Dynamics (UED) offers a different take. What if there's no need to curve a geometric "fabric"? What if the very concept of fabric is misleading?

In UED, there is no spacetime to curve—only a structured vacuum field (SVE), rich with equilibrium energy. When mass concentrates in a region, it doesn't warp geometry—it changes the ripple tension in that zone. This creates a gradient—like a slope in the ripple field. Objects don't fall *because* of geometry. They fall *into* alignment with the local ripple slope—a passive drift toward lower field tension.

It's not a pull. It's not a push. It's not even an acceleration in the classical sense. It's a realignment of resonance.

The Paradigm Shift

UED invites us to ask: Was Einstein describing the effect of gravity—while missing its cause? Did we mistake the appearance of motion in curved coordinates for the reality of ripple adjustment in a structured field?

This section reopens the most sacred assumption in modern physics—that space curves—and dares to suggest: perhaps we bent the map, not the terrain.

The Elevator Thought Experiment Reframed

Albert Einstein's elevator thought experiment is one of the most elegant mental models in the history of physics. It was his way of stepping outside Newton's box—and into the curvature of spacetime.

Let's revisit that elevator—but this time, not with Einstein's eyes, but through the ripple lens of Universal Energy Dynamics (UED).

Einstein's Elevator: Gravity as Indistinguishable from Acceleration

Einstein imagined a person inside a sealed elevator:

- If the elevator is in deep space and accelerating upward, the person feels pressed to the floor.
- If the elevator is stationary on Earth, the same feeling occurs due to gravity.

His conclusion? There is no way to distinguish acceleration from gravitational pull—they are equivalent. This principle of equivalence led Einstein to propose that gravity wasn't a force but a property of spacetime geometry. The floor wasn't pushing up—the elevator was following a curved geodesic path in a warped spacetime.

This reasoning helped usher in General Relativity. But it also carried a hidden assumption: that curvature causes motion. That geometry explains experience.

The UED Correction: Geometry Describes It—Ripple Alignment Causes It

In the UED model, we flip the assumption. Instead of interpreting free-fall as a *trajectory in curved space*, we see it as a passive realignment through ripple gradients in Structured Vacuum Energy (SVE).

From this perspective:

- The elevator isn't accelerating in space—it's sinking into a ripple slope, like a marble rolling down a smooth energy gradient.
- There's no external force acting. The "fall" is a consequence of the elevator's field structure adjusting to a lower-tension zone.

This is not acceleration in the Newtonian sense. Nor is it motion through geometric warping. It's something subtler and more elegant: A resonance drift toward balance.

Free-Fall Is Not Force-Free — It's Friction-Free

UED reframes free-fall as not the absence of force, but the absence of resistance:

- There is no push or pull.
- The elevator isn't being acted upon.
- Instead, it is naturally moving along a ripple pathway—the

path of least ripple tension, or minimum field resistance.

The body inside feels weightless not because there's no force, but because its field is no longer resisting. It's harmonized. It's aligned.
That's why astronauts in orbit experience "microgravity"—not because gravity has disappeared, but because they're riding the ripple, perfectly aligned with the gradient. There's no resistance left to feel.

Insight: Acceleration Is an Artifact of Misalignment

In conventional physics, acceleration is a change in velocity. In UED, acceleration is the field's attempt to realign with a shifting gradient.
So when you fall:

- You don't speed up in space.
- You reduce your misalignment with the ripple field.
- The closer you are to perfect alignment, the less effort your field exerts—and the more effortless your motion feels.

It's not force-less—it's friction-less.

Reframing Einstein's Genius

Einstein's elevator revealed something profound: the experience of gravity and acceleration are experientially identical. But perhaps that's because they are both just field responses to ripple tension. Not two different phenomena—just two different angles of alignment.
UED doesn't discard Einstein's insight. It completes it.
By removing the need for geometry and returning to ripple reality, we now see:
"What falls does not fall *down*. It falls *in* — into perfect resonance."

Ripple Gradients – The New Field Topology

Classical physics explained gravity as a force. Relativity replaced that with geometry. But both frameworks share a common flaw: they treat gravity as something that acts.

Universal Energy Dynamics (UED) reframes gravity not as something *doing* anything — but as something *being* something. A ripple slope. A gradient. A tension map in the Structured Vacuum Energy (SVE) field.

This is not a metaphor. It's a literal ontological shift.

SVE as a Ripple-Tension Field

Structured Vacuum Energy is not empty space.

It is a dense, structured, and ultra-responsive field that holds ripples, nodes, and alignment zones. Every so-called "object" is simply a stabilized ripple structure within this field.

And crucially — this field has tension.

Just like a stretched rubber sheet or a surface of water, variations in ripple density across SVE create slopes. Not in space. In energy configuration. In resonance.

And those slopes are what we perceive as "gravity."

Motion Follows the Gradient

In UED, objects don't "fall" because of attraction or force. They descend ripple gradients, naturally seeking zones of minimum tension and harmonic equilibrium.

This is the same principle that governs:

- Water flowing downhill (due to gravitational potential)
- Heat spreading across a cold surface (due to temperature gradients)
- Sound moving from loud to quiet zones (due to pressure differences)

Gravity, in this model, is simply the tension topology of the SVE field.

There is no "pull." There is only passive descent into lower ripple strain.

The Pull Illusion

So why do we feel pulled down? Why does an apple "fall" from a tree?

Because the observer's body (their field) is misaligned with the gradient—and it realigns through motion. The path to harmony feels like movement, just as flowing water *appears* to be pulled by Earth. But no one asks, "Who is pulling the water?"

We're conditioned to believe in invisible hands.

But UED says: there are no hands — just slopes.

Just as a marble rolls down a curved surface, a massed ripple node rolls down the ripple slope — effortlessly, without needing a push or a pull.

Why Curved Spacetime Becomes Unnecessary

General Relativity imagines massive objects curving spacetime, which tells smaller objects how to move.

But what if curvature itself is a misinterpretation?

From the UED view:

- Mass doesn't curve geometry.
- Mass creates ripple strain in SVE.
- That strain forms gradients.
- Objects align along those gradients — not because of attraction, but because of field tension seeking equilibrium.

In other words, the curvature Einstein described is not wrong — it's a geometric rendering of ripple tension, not a fundamental structure of space.

We don't need a rubber sheet metaphor.

We just need to accept:

"What we called gravity was never geometry — it was gradient."

Final Insight:

This section changes the gravitational game entirely. No more mysterious force. No more imaginary curves.

Just a field. Just structure. Just resonance following slope.

The Mass–Tension Coupling

At the heart of Universal Energy Dynamics (UED) lies a radical yet elegant idea: Mass is not an object—it is a consequence. A ripple consequence. A signature of field tension within the Structured Vacuum Energy (SVE).

Let's unpack this redefinition and how it transforms our understanding of gravity and interaction.

Mass = Localized Field Tension

In classical physics, mass is a "thing" with substance.

In General Relativity, mass becomes a source of curvature.

But in UED, mass is the condensation of ripple tension into a stable node — a point of internal harmony sustained by stress in the surrounding field.

You don't "have" mass. You are a tension pattern held steady within the ripple fabric of the vacuum.

- A rock is not pulling on you.
- It's simply shaping the field around it, like a heavy object distorting a trampoline—not by weight, but by ripple density.

Mass, therefore, is a field topology, not a unit of matter.

More Mass = Steeper Ripple Slope

The greater the field tension localized in an area, the steeper the ripple gradient around it.

This is not an external force. It's a natural consequence of equilibrium-seeking behavior. Just as:

- A stretched sheet pulls neighboring points toward a dent.
- A hot surface influences nearby molecules.
- A loud sound spreads pressure across air.

Likewise, a high-tension ripple node warps the SVE around it, producing a slope that other nearby ripple fields naturally realign to. The effect we call gravitational attraction is really ripple descent — a passive alignment with the gradient of lowest ripple resistance.

Not Attraction, But Field Shaping

In the old model, objects "pull" on each other.

In UED:

Bodies do not attract — they shape.

Each mass imprints its ripple geometry onto the surrounding field.

That geometry doesn't push or pull — it simply *exists* as an energetic contour.

Other bodies entering that region begin to realign themselves to that contour — not due to force, but due to resonance minimization.

It's not different from how:

- A guitar string resonates when exposed to the right pitch.
- A tuning fork starts to vibrate when brought near another of the same frequency.

Alignment is not forced — it is enabled by matching conditions.

The Consequence Chain

Here's the ripple logic:

1. Mass forms from stabilized ripple tension.
2. That tension warps the surrounding SVE into a slope.
3. Nearby ripple fields enter that slope and begin realigning passively.
4. The realignment appears as motion — what we traditionally call "falling" or "attraction."

Nothing was pulled. No forces were exchanged. Just intelligent feedback through ripple geometry.

Key UED Insight:

"Mass doesn't pull. It tells the field how to behave—and the field tells others where to go."

This section demystifies the invisible hand of gravity. It's not invisible because it's magical—it's invisible because it's structural.

Why Orbits Work Without Geometry

For centuries, we've imagined celestial bodies as being "caught" in the grip of invisible curves—be it Newton's gravity pulling them in, or Einstein's spacetime geometry shaping their path.

But what if orbits don't depend on force or curvature at all?

What if they are simply ripple equilibrium loops—stable feedback patterns sustained by the field dynamics of Structured Vacuum Energy (SVE)?

The Old Story: Curved Space, Curved Paths

In classical mechanics:

- Planets are in constant free-fall, veering around a center due to a gravitational force.

In General Relativity:

- Gravity is not a force but the bending of spacetime, and objects follow "geodesics"—the shortest path in curved geometry.

These views differ in language, but both assume motion through space, shaped by an external geometry.

But both still treat the object as traveling.

The UED Reframe: Orbit as a Stable Ripple Path

Under Universal Energy Dynamics (UED), this entire framing is inverted.

The planet is not a traveler—it is a field node co-resonating within a larger ripple slope.

It's not "orbiting" in the traditional sense. It's locked into a ripple rhythm, just like a leaf floating in a gentle whirlpool, not pulled—but held by feedback loops.

There is no trajectory—only resonance maintenance in a curved ripple topology.

The "path" of the orbit is a side effect of the field alignment needed to maintain equilibrium between the planet's own ripple signature and that of the Sun's massive ripple basin.

Surfing, Not Falling

We typically imagine Earth as "falling toward the Sun and missing"—an endless dance of gravity and motion.

But UED proposes a more elegant metaphor:

- The Earth isn't falling.
- It's surfing a ripple path—gliding along the contours of a pre-existing field pattern in SVE.

This field isn't curving space—it's layering tension in dynamic gradients. Where that tension achieves balance, you get orbital stability.

Where tension is uneven or disturbed, you get drift, spirals, or escape.

Lagrange Points: Ripple Balance Zones

Nowhere is this redefinition clearer than at the Lagrange points—locations in a two-body system (like Earth and Sun) where gravitational "forces" seem to cancel.

In traditional mechanics, they're points where the net gravitational vector and centrifugal effects balance out.

But that's just math describing outcome.

In UED, Lagrange points are simply field-neutral zones—locations where ripple gradients overlap in a way that produces minimum misalignment for an object to remain relatively still with minimal effort.

These are resonant pause-points, not tricks of math.

Orbits Without Geodesics

Let's state this plainly:

- You don't need curved space to explain orbits.
- You don't need gravitational "force" to sustain planetary motion.
- What you need is ripple harmony—the stability of a system within a local SVE topology.

The Earth doesn't swing around the Sun because of invisible geometry—it stays aligned with the field slope, tracing an energetic "groove" that's been carved by feedback and balance.

The groove is not in space—it is in structure.

UED Summary:

"An orbit is not a motion around a point. It's a dance in a ripple—where resistance vanishes, and resonance flows."

Black Holes and the Ripple Abyss

Black holes have long haunted the edge of physics—a place where gravity allegedly crushes matter into singularities and space folds into itself. But what if we've misunderstood them entirely?

Under Universal Energy Dynamics (UED), black holes are not geometrical monsters hiding a mathematical absurdity. They are high-density ripple zones, where field tension reaches its most extreme state of misalignment resistance.

And that changes everything.

No Need for Singularities or Infinite Curvature

In General Relativity (GR), black holes emerge from the math as singularities—points where spacetime curvature becomes infinite, and the laws of physics break down.

This is not an explanation—it's a red flag.

Any theory that generates infinities is a clue that the foundational assumptions are incomplete or flawed. GR, despite its elegance, lacks a physical substrate. It predicts the behavior of objects as if geometry alone were causal—but ignores the medium through which these behaviors manifest.

UED provides that missing piece: Structured Vacuum Energy (SVE).

Within this framework, a black hole is not a "hole" in space—it's a zone of extreme ripple compression, where the SVE field has condensed into maximum tension alignment.

Ripple Abyss: A Field Without Exit

Imagine the ripple field as a stretched membrane.

- A star is a localized tension node.
- A neutron star is a sharper distortion.
- But a black hole is where the membrane pinches so tightly that no outward ripple reconfiguration is possible.

There is no "pull"—just an inward gradient so steep that all ripple interactions become trapped, unable to realign back to external observers.

This isn't about mass bending geometry. It's about ripple slope exceeding alignment threshold—so perception itself becomes impossible from the outside.

Why Light "Bends" Near a Black Hole

Light does not "bend" because of spacetime curvature. That's a visual metaphor, not a mechanistic explanation.

Under UED:

- Light is a ripple alignment event between source and observer.
- Near a black hole, the local ripple field is so steep, so distorted, that the path of least resistance is dramatically

curved.

- From our field's perspective, the light seems to bend—but really, we are perceiving alignment distortion, not photon travel.

No particle is pulled. No geometry is warped. Instead, the structure of potential alignment itself has changed.

The Event Horizon as ATT Barrier

The event horizon is often described as the "point of no return." But why?

UED provides a clean reframe:

The event horizon is where the Alignment Threshold Time (ATT) becomes functionally infinite.

This means:

- Any attempt by an external ripple field to align with the black hole interior exceeds the observer's perceptual threshold.
- Not because something is "traveling slower" or "stuck," but because alignment cannot occur within a coherent timeframe for perception to happen.

That's why we say nothing escapes—not because it's swallowed, but because the feedback loop breaks.

From the black hole's side, all is internally realigned. From our side, all signals vanish into unresolvable ripple resistance.

Closing Insight: A Black Hole Is Not an Anomaly—It's a Ripple Limit

Black holes are not tears in space or containers of infinite density.

They are:

- Boundaries where SVE tension exceeds external feedback capacity.

- Regions where alignment fails—not because nature breaks, but because perception stalls.
- Proof that reality is governed not by geometric absurdities—but by ripple balance, resistance, and the threshold of awareness.

"The black hole is not where light dies. It's where your ripple can no longer meet its reflection."

The End of Force-Based Gravity

For centuries, gravity has been treated as a force. From Newton's falling apple to Einstein's spacetime curvature, the assumption of action has remained consistent—something causes gravity to "act" on objects. But what if this very assumption is the flaw?

Universal Energy Dynamics (UED) dismantles the notion of gravity as a pushing or pulling agent and replaces it with a ripple field consequence. In this framework, gravity doesn't act at all—it responds to imbalance.

Force, as Classically Defined, Disappears in Ripple Logic

In Newtonian mechanics:

- Force is what causes acceleration.
- Gravity is an attractive force between masses.
- Equations like $F = G(m_1m_2)/r^2$ give the illusion of a measurable "cause" behind the attraction.

In General Relativity:

- Gravity is no longer a force, but an emergent effect from geometry.
- Mass tells space how to curve; space tells mass how to move.

While this shift was revolutionary, it still depends on a reactionary model—a deformation (of geometry) that produces behavior (acceleration).

UED goes a step further. It reinterprets everything in terms of ripple alignment and tension gradients.

In this new view:

- There is no *force* causing motion.
- There is no external *geometry* commanding behavior.
- There is only the inherent tendency of ripples to resolve misalignment and find lower resistance paths.

Gravity vanishes as a *thing*, and becomes a field adjustment—not unlike how a stretched fabric naturally settles back into equilibrium when disturbed.

Gravity = Field Consequence, Not External Imposition

Gravity, in UED, is not imposed upon objects—it is emitted by field configuration.

- When a massive ripple node (like a star or planet) condenses tension in the Structured Vacuum Energy (SVE), it reconfigures the surrounding ripple field.
- Other nearby ripple nodes (objects) do not experience a force pulling them—they realign to this local tension gradient.
- What appears as falling, accelerating, or curving paths are actually passive field responses to the ripple terrain.

This makes gravity similar to buoyancy or thermal equilibrium—an outcome of relative states, not an external agent with intent.

No "Graviton" Needed — Ripple Alignment Is Continuous and Non-Quantized

The search for the graviton—a hypothetical quantum particle that mediates gravity—has consumed decades of theoretical physics.
But what if there is nothing to find?
UED challenges this pursuit by asserting:

- Gravity is not quantized like electromagnetic or nuclear forces.
- It is a continuous realignment process within a tensioned ripple field.
- No particle exchanges occur. Instead, tension harmonizes across distances based on ripple compatibility.

In short, gravity has no messenger because it has no message. It is not sent. It is felt—as immediate mutual realignment between field structures.
This elegant shift simplifies gravity from an interaction into a relation. Not a force. Not a particle. Just shared structure seeking coherence.

Closing Thought: "Gravity Doesn't Act. It Responds."

This final insight reframes everything we thought we knew:
Gravity doesn't pull. It doesn't curve. It doesn't act. It responds—to imbalance, misalignment, and field tension.
And in doing so, it reveals that the universe is not being acted upon, but constantly realigning with itself.

Closing Reflection

"You're not being pulled down. You're being invited in—by the shape of the field beneath your feet."

We've been taught to see gravity as a force—an invisible agent pulling us downward, yanking planets into orbits, bending light around stars. The story was compelling. Predictive. Functional. But perhaps it was never complete.

The force of gravity, as we were told, was a command—an undeniable directive from mass to motion. Yet what if it's something far subtler... and far more profound?

What if you're not being pulled at all? What if you're simply responding to an invitation—a field whisper beneath your feet, shaped by tension you didn't create, but now participate in?

In the UED framework, gravity is no longer a battle against forces or curves. It is the natural surrender of your ripple field to the path of least resistance—like a leaf finding the quietest part of the stream.

You are not at the mercy of a universal tug. You are in dialogue with the structure of existence itself.

This closing thought reframes not only motion—but meaning. Because when gravity stops acting on you and begins resonating with you, the universe is no longer external. It's intimate. Alive. Interactive.

And that makes every fall a form of participation—not defeat.

Part III: Completing the Equation

Chapter 10: Where Relativity Breaks

Dark matter, dark energy, singularities—why GR stumbles and stalls

"A great theory doesn't need to explain everything—but it must not invent what it cannot explain."

They applauded it as the most elegant theory of the 20th century—a revolution in thought, a triumph of imagination over limitation. Einstein's General Relativity redefined gravity not as a force but as geometry, and the world never looked back.

But the universe did.

Because beyond the applause lay a series of stubborn riddles—galaxies spinning too fast, the cosmos expanding too strangely, and equations collapsing into infinities. In each case, the theory didn't evolve—it evaded. And what it couldn't explain, it invented: dark matter, dark energy, inflation, and singularities. Ghosts created not by observation, but by desperation.

This chapter is where the honeymoon ends.

We're not here to dismiss Einstein. We're here to show what happens when a theory forgets the field it stands on. When it clings to curvature even when reality ripples. When its math runs ahead of its meaning.

Welcome to the limits of relativity. Not its failures—but the consequences of its refusal to change.

The Cracks Behind the Applause

Every great myth begins with a miracle—and in modern physics, that miracle was General Relativity.

Einstein's equations didn't just predict the bending of starlight—they *explained* it, down to the decimal, during a total solar eclipse in 1919. The orbit of Mercury, long a thorn in Newton's side, suddenly made sense. And in our own era, satellites circling Earth account for relativistic time dilation to keep GPS precise—down to centimeters. These were not minor victories. These were scientific showstoppers. Relativity had not only entered the textbooks—it had conquered them.

But applause can be deceiving.

A theory's brilliance in one domain doesn't guarantee truth in another. Newton explained cannonballs and comets—but couldn't fathom Mercury. Relativity explained Mercury—but stumbled when it looked out at galaxies and back toward the Big Bang.

Because beyond the solar system, things stop behaving as expected.

- Galaxies rotate too fast, as if held together by invisible glue.
- The universe is expanding, but not at a constant rate—it's accelerating.
- Black holes don't just bend light—they break equations.
- The beginning of time, as described by GR, ends in a singularity—a place where logic vanishes.

And how did mainstream physics respond? Not by fixing the foundation, but by patching the cracks with undetectable particles, unverifiable fields, and infinities that must be true because the math says so.

At some point, we must ask:

What happens when the map—however beautiful—no longer fits the terrain?

Because a map that invents landmarks to justify itself is no longer a map. It's fiction.

The cracks behind General Relativity's applause aren't signs of failure—they're signs of a deeper truth waiting to be seen. A truth not curved in space—but rippling through reality.

The Problem with Dark Matter

The trouble began in the outskirts of galaxies.

Astronomers expected stars farther from the galactic center to orbit more slowly—just like planets farther from the Sun. That's how gravity is supposed to work. The farther you go, the weaker the pull. But when they measured the speeds, something strange emerged: those distant stars weren't slowing down—they were whipping around at nearly the same speed as those closer in.

According to Newton—and even Einstein—this shouldn't happen. The gravitational force from all visible matter—stars, gas, dust—simply wasn't enough to hold galaxies together at those speeds. The stars should be flung into the void. But they weren't. They were locked in tight spirals, like clockwork that refused to obey the clock.

So, what did science do?

It invented an invisible glue: dark matter.

A mysterious, undetectable form of mass. It doesn't emit light. Doesn't reflect it. Doesn't absorb it. It passes through us, through planets, through everything—and yet it supposedly makes up 85% of the matter in the universe.

Why?

Because General Relativity couldn't explain the motion.

Instead of questioning the foundation—physicists built a scaffolding of speculation. They didn't find new matter—they imagined it, to save the theory that once saved Newton. It became a new religion: If we can't see it but need it, it must be there. Observation bowed to preservation.

But what if the fault lies not in the matter—but in the model?

What if galaxies aren't disobeying gravity—but revealing that gravity isn't what we thought?

Enter UED: The Ripple View

In Universal Energy Dynamics, matter is not the source of motion—it's a consequence of ripple alignment. Stars orbit in ripple inertia zones, where structured vacuum energy (SVE) creates dynamic equilibrium fields. What we interpret as gravity is just the tension gradient of ripples, not the force from some central mass.

From this view:

- The flat rotation curves aren't defying gravity.
- They're tracing harmonic ripple loops maintained by SVE.
- The outer stars aren't being held in—they're riding the rhythm.

No need for invisible mass. No need for dark matter detectors that never detect. Just a deeper understanding of the field beneath our feet—and the ripples that carry the sky.

Because maybe the universe isn't missing mass. Maybe physics is missing meaning.

The Mirage of Dark Energy

Just when physicists were settling into the mystery of missing mass, the universe delivered another shock.

The cosmos, they had believed, was expanding—but slowing down. After all, if gravity pulls things together, the collective mass of all galaxies should gradually put the brakes on expansion. That was the story for decades.

Until it wasn't.

In the late 1990s, two separate teams of astronomers observing distant supernovae made a startling discovery: the expansion of the universe wasn't slowing down—it was speeding up.

The further out they looked, the faster things seemed to be receding.

This was not a tweak—it was a reversal. It didn't just challenge relativity—it confounded it.

So once again, a fix was summoned from the void.

Physicists reached for a forgotten relic: Einstein's cosmological constant (Λ)—a term he had originally inserted into his equations to keep the universe static, only to later abandon and call his "greatest blunder."

Now, Λ was reborn—not to hold the universe still, but to push it apart.

But what is this force? Where does it come from? What governs its strength? No one knew. So they gave it a new name: dark energy—a placeholder for ignorance dressed as discovery.

It now accounts for about 70% of the universe's energy content. Not because we understand it—but because we can't explain the acceleration without it.

Let that sink in: More than two-thirds of the known universe is assigned to something we can't detect, measure, or define—just to make the equations work.

UED: The Field Doesn't Push—It Realigns

In Universal Energy Dynamics, there's no need to imagine an external force causing expansion.

The universe isn't accelerating *through space*—it's reconfiguring its field. The ripple topologies of structured vacuum energy (SVE) are not static—they realign to restore equilibrium after cosmic disturbances, such as massive matter formations, supernova bursts, and early turbulence post-origin.

From this view:

- Expansion is not driven—it's resolved.
- The so-called "acceleration" is a ripple gradient smoothing process, not a cosmological propulsion.
- There's no dark energy—just a misreading of ripple

evolution.

It's like watching the surface of a pond settle after a rock has splashed. The ripples move outward—not because something pushes them—but because the medium seeks stillness.

We mistook this natural ripple balancing as cosmic acceleration. We called it energy—but it was intelligence in motion.

Because sometimes the universe doesn't need more energy. It just needs less misunderstanding.

Singularity — The Mathematical Panic Button

When all else fails, divide by zero.

That's not written in any physics textbook—but it might as well be, given how modern cosmology handles its biggest questions.

At the heart of two of the most dramatic ideas in modern physics—the Big Bang and black holes—lies the same uneasy truth: General Relativity breaks down completely.

It doesn't bend, it doesn't flex—it snaps.

When equations are pushed to the extreme, GR doesn't offer insight. It offers surrender. The math leads to what's called a singularity—a point of infinite density, zero volume, and undefined behavior.

In the Big Bang scenario, the entire universe is imagined to have emerged from such a singularity: a timeless, formless point where physics no longer applies.

In the case of black holes, once enough mass is crammed into a small enough space, gravity becomes so intense that even light cannot escape—and deep within, the same fate awaits: a singularity of infinite collapse.

But let's be clear—a singularity isn't a discovery. It's a white flag.

It doesn't describe reality. It admits that our tools can no longer describe anything at all.

Infinite density is not a physical condition—it's a mathematical malfunction. It's the moment the equation throws its hands in the air and says, "I don't know what happens next."

And physics, as a discipline, quietly walks away.

UED: There Are No Singularities—Only Misread Fields

In the Universal Energy Dynamics framework, there is no need for infinities, because there is no such thing as an absolute collapse of space or time.

Instead of space bending endlessly inward, ripple tension within Structured Vacuum Energy (SVE) increases toward a threshold of alignment breakdown. What appears as a "collapse" is actually an extreme compaction of ripple fields, where local equilibrium becomes highly distorted—but never broken beyond repair.

In black holes, for example:

- The center isn't a point of infinite density—it's a zone of ripple saturation, where field tension reaches its maximum sustainable limit.
- Beyond that limit, perception fails—not because reality ends, but because ripple realignment becomes functionally impossible for any external observer.

In the case of the early universe:

- UED doesn't suggest a bang from an absolute point.
- It sees an emergence from ripple compression, a reconfiguration within the SVE field, where time, motion, and matter phase in through structured alignment—not explode out from a void.

There is no panic. No math-breaking chaos. No "beginning" where laws cease to exist.

Just an evolving field. A tension. A ripple. A consequence playing out through cosmic harmony—until misread as madness.

Singularities don't exist in the universe. They only exist in theories that couldn't handle the universe.

The Inflation Patch

If the Big Bang was the explosion, cosmic inflation was the duct tape. The early universe, as described by General Relativity, wasn't behaving nicely. After the supposed "singularity," the universe should've exploded outward chaotically—but observations showed something odd: the universe was surprisingly smooth. Too smooth.

The cosmic microwave background (CMB)—a faint afterglow of the early universe—looked nearly uniform in all directions. That shouldn't happen. If the Big Bang was a wild eruption, why didn't it leave jagged temperature scars across the sky?

Physicists needed a fix. So they reached again—not for observation, but for imagination.

They proposed a moment of inflation—a hyper-exponential expansion of the universe in a tiny fraction of a second, far faster than the speed of light.

According to this patch:

- Inflation stretched space itself by a factor of at least 10^{60}.
- It smoothed out wrinkles and flattened the geometry.
- And then—mysteriously—it stopped, just in time for regular Big Bang physics to take over.

To make it work, physicists introduced a hypothetical scalar field—the inflaton—that drove inflation and then conveniently decayed away.

But what is the inflaton? Where did it come from? Why did it switch off so perfectly?

No one knows.

In fact, no one has ever detected it, or even explained its properties in a consistent way. It exists because the math needed it—not because reality revealed it.

Once again, the equation led—and the universe was expected to follow.

UED: Ripple Fields Don't Inflate—They Harmonize

Universal Energy Dynamics sees the early universe not as an explosion needing cleanup—but as a ripple field finding its shape.

After the initial ripple ignition within Structured Vacuum Energy (SVE), the universe didn't explode chaotically—it resonated.

- Early ripple disturbances were intense, but followed patterns.
- Local misalignments created energy density variations—what we perceive today as fluctuations in the CMB.
- Over time, these distortions settled through harmonic ripple resonance, creating natural zones of coherence and balance.

From this view:

- The "smoothness" of the universe is not because of inflation—but because ripples harmonize.
- The large-scale uniformity is the signature of ripple equilibrium, not artificial expansion.
- No scalar fields. No phantom energies. No switch-off problems.

Just field behavior under pressure, working toward realignment—exactly as every ripple does, from a puddle to a planet.

Inflation didn't explain the universe. It explained a theory's discomfort with what the universe already knew how to do: Balance itself.

Too Many Fixes, Too Little Foundation

Every theory faces challenges. That's natural. But when a theory responds to every challenge with a new invention instead of a deeper understanding, it's not evolving—it's crumbling under its own weight.

That's where modern cosmology stands today.

Let's pause and take inventory:

- **Dark Matter**: Proposed to explain why galaxies don't fly apart despite not having enough visible mass. ➤ Undetectable. Unconfirmed. Still essential—because gravity as we know it can't hold galaxies together without it.
- **Dark Energy**: Invented to explain the accelerating expansion of the universe. ➤ No known source. No lab detection. Yet it's now said to comprise 70% of the universe's energy content.
- **Cosmic Inflation**: A lightning-fast expansion post–Big Bang, introduced to smooth out early irregularities. ➤ Driven by a scalar field we've never observed, using rules we don't fully understand, that ends at exactly the right time—because it must.
- **Singularities**: The endpoints of black holes and the origin of the Big Bang, where equations collapse into infinities. ➤ Not physics—just the limit of flawed math pretending to describe reality.

None of these are *derived from first principles*. None of them come from observed causes. Each one was inserted because General Relativity—brilliant as it is—could not explain what we were seeing.

It's like trying to patch a leaky boat with duct tape, chewing gum, and faith. Each fix buys time, but none fix the boat. At some point, you stop patching and ask:

Was the hull ever seaworthy?

UED Offers No Crutches—Only Coherence

Universal Energy Dynamics doesn't need exotic substances or imagined energies. It doesn't rescue failing predictions with invisible actors. It rewrites the script.

- **Galactic behavior?** ➤ Explained through ripple inertia in structured vacuum equilibrium.
- **Cosmic expansion?** ➤ Understood as ripple field realignment—not outward motion but field adjustment.
- **Black holes?** ➤ Not infinite collapses, but maximum alignment tension zones within SVE.
- **Cosmic origin?** ➤ Not a singularity, but a phase transition within the ripple medium.

Where GR adds patches, UED seeks patterns. Where GR preserves the geometry, UED listens to the rhythm.

Because the problem was never the observations. The problem was what we thought the universe *should be*—instead of asking what it *actually is*.

The Ripple Advantage: Replacing the Fixes

Let's imagine a different kind of science.

One that doesn't rush to invent invisible particles every time the numbers don't add up. One that doesn't force the universe to obey its equations, but listens to the universe first—and then writes the math. One that doesn't bend space or invent "dark" substances but starts with the most fundamental medium of all: Structured Vacuum Energy (SVE).

That's where Universal Energy Dynamics (UED) begins. Not with matter. Not with force. Not even with spacetime. But with the ripple—the primary mode of expression for energy in action within a structured field.

While General Relativity works with curvature, coordinates, and geometry, UED deals with alignment, tension, and consequence. And with this shift in lens, many of the so-called "unsolvable" problems in physics become not only solvable—but natural.

Let's revisit the four biggest anomalies—now through the ripple view:

Galactic Rotation = Ripple Inertia Zones

In UED, stars don't need extra mass to stay bound to galaxies. Their motion is governed by ripple harmonics within the SVE field. These stars are riding the inertia zones—like eddies in a whirlpool—not being yanked inward by gravitational force. There's no dark matter required. Just misunderstood ripple dynamics.

Cosmic Expansion = Realignment, Not Acceleration

The universe isn't speeding outward like a rocket ship. It's reconfiguring from within. Ripple fields disturbed during major cosmological transitions settle over time, leading to what looks like expansion—but is actually a shift in field structure, not in distance. No need for dark energy. Just a more accurate perception of what "motion" means when the medium itself is in flux.

Black Holes = Tension Fields, Not Bottomless Pits

Black holes in UED aren't holes in spacetime—they're zones of maximum ripple compression. What we call the "event horizon" is simply the threshold where ripple alignment fails—where ATT (Alignment Threshold Time) becomes functionally infinite. Beyond that, perception collapses—not reality. There's no singularity. No division by zero. Just the limits of observational alignment.

The Background Field Isn't Empty—It's Misread

All of this becomes possible because UED doesn't treat the vacuum as "nothing." It treats it as everything—a structured, energy-rich medium that holds, conducts, and balances all interaction. Where GR needs entities to fix gaps, UED offers interpretation to reveal order.

No ghostly particles. No infinite densities. No fields from nowhere that start and stop by mathematical convenience.

Just a field that speaks in ripples. And a science willing to listen.

Because when you understand the rhythm of the medium, the music of the cosmos finally makes sense.

Closing Reflection

"Relativity broke not in its math—but in its mind."

General Relativity was not a failure of intelligence—it was a triumph of imagination. It dared to see gravity not as a force, but as form. It gave geometry a starring role in the theater of the cosmos.

But even the most elegant stage collapses when the script no longer fits reality.

Relativity broke not because its numbers didn't match—it often got them right. It broke because **it refused to change the frame** when the picture blurred. It held tightly to **curvature, coordinates, and constants**—even when the universe was asking different questions.

It created **patches**, not patterns. It invented **entities**, not understanding. It answered observations not with openness, but with **explanations by necessity**—dark matter, dark energy, inflation, singularities.

Yet the real flaw wasn't the math. It was the mindset.

UED does not discard Einstein—it evolves the inquiry.

It shifts the question from "What shape is space?" to **"What tension lives in the field beneath it?"** It asks not "What mass is missing?" but **"What alignment is misunderstood?"** It doesn't warp the map to fit the cracks in the road. It redraws the landscape from ripple up.

Because reality was never built on curves. Not even on matter. It was built on **consequences**—the ripple effect of energy in action through a structured field.

And when you finally stop bending your equations to preserve the past, You begin to perceive the **field that was always there**— Stable, structured, singing in silence, waiting to be understood.

General Relativity helped us see. Universal Energy Dynamics helps us hear.

Chapter 11: From Frames to Fields

Why UED doesn't need observers in motion—it focuses on ripple logic
"If the laws of nature change with your frame, then maybe the frame was never real."
For over a century, we've been told that motion is a matter of perspective. That time bends differently for different travelers. That two observers watching the same event from different speeds or directions may not agree on what happened—or even *when* it happened.
This was Einstein's great insight: that there is no absolute rest, only motion relative to a chosen frame. And for a while, it worked. Relativity helped us calculate. It helped us predict. But it also buried something deeper beneath layers of shifting coordinates and perceptual contradictions.
Because if reality itself is conditional—if truth is always frame-dependent—then what anchors existence at all? Are we really to believe that the universe rewrites its rules depending on who's watching?
Or have we confused the surface of perception with the substance of reality?
Universal Energy Dynamics doesn't ask what happens from your point of view. It asks: What happens to the field—regardless of who's watching?
This chapter marks the break from relativity's observer theater. We step off the train, out of the elevator, and beyond the twin paradox—into the quiet logic of ripples.

The Relativity of Observers

Einstein's theory of relativity did something radical—it removed the idea of absolute motion from physics altogether.

No longer was there a privileged "rest frame" of the universe. Whether you're sitting on a train, standing on the platform, or flying past in a spaceship, your motion is as valid as anyone else's. All motion is relative—that was the foundational idea.

In this view, there is no universal stillness, no cosmic reference point. Instead, reality is described relative to an observer's frame of reference.

These frames come in two types:

- Inertial frames, which move at constant velocity (no acceleration), where Newton's first law holds.
- Non-inertial frames, which involve acceleration and require fictitious forces (like centrifugal force) to make sense of motion.

Einstein's genius was in showing how even gravity could be seen as acceleration in a non-inertial frame, giving rise to the idea that gravity is not a force, but the bending of spacetime itself.

But hidden in this innovation was a subtle shift—one that would eventually corner physics into a conceptual trap.

Observers Became the Architects of Reality

By anchoring physics in frames of reference, the observer became central to the story. What you experience depends not on some universal event, but on how fast you're moving, where you are, and how your clock ticks relative to others.

This leads to bizarre but mathematically valid consequences:

- Two observers disagreeing on the order of events.
- Clocks ticking at different rates based on velocity.

- Rulers shrinking or stretching depending on motion.

From a computational perspective, this works. From a philosophical perspective—it unravels.

Because if reality itself depends on who's watching, then is there anything that exists independently at all? Do events only become real when perceived from a frame? Can nature be defined without an observer?

Relativity, despite its beauty, left us with an uncomfortable implication: Truth is no longer intrinsic—it's negotiated between perspectives.

But perhaps that was never a truth—only a transitional tool. Maybe the universe isn't observer-dependent after all. Maybe the field was there the whole time—quiet, stable, and rippling—regardless of how fast you were moving or how your clock ticked.

That's where UED enters: not to deny what relativity got right, but to free physics from frames and fictional forces, and return to what's real.

The Hidden Cost of Frames

Relativity taught us to measure from wherever we stand. But in doing so, it quietly changed the meaning of measurement itself.

Once you say that all motion is relative, every observation becomes relational—not intrinsic. The value you measure depends on where you are and how fast you're moving. There is no absolute time. No universal length. No shared simultaneity. Everything bends around your frame.

And that sounds clever—until you ask,

"Then what is real?"

Let's take two of relativity's most famous effects:

Time Dilation

Two clocks, moving at different speeds, tick at different rates. One astronaut flies away at near-light speed, returns decades later, and finds their twin has aged much more. But did time itself change? Or is this just the consequence of how we measure time under distortion?

Length Contraction

An object moving close to light speed appears shortened in the direction of motion. But shortened to whom? From the object's perspective, nothing changes. So is the contraction real—or a perspective illusion rooted in how frames deform space?

Relativity allows both to be "true" depending on your frame. But physics, at its core, is meant to uncover what's true regardless of where or how you look.

That's where the hidden cost shows up.

Relativity's brilliance comes at the price of ontological clarity. By making every truth frame-dependent, it sacrifices the idea of a singular, causally grounded universe.

If ten observers can each claim different durations, distances, and even event orders—and all are equally valid—then what holds the story together?

It becomes difficult to speak of causality in any unified way. Did A cause B? Or did B happen before A in another frame? The line between sequence and simultaneity becomes blurred.

This leads to what we can call multiplicity of truths:

- A world with many clocks, none absolute.
- Many versions of the same event.
- Many geometries warping based on who is observing.

But a multiplicity of truths without a common field of consequence risks becoming chaos dressed in coordinates.

The universe doesn't behave like that. Stars don't go flying off because someone watches them differently. Atoms don't decay based on belief. And galaxies don't rotate in different ways for different frames.

What's needed isn't more frames. It's a deeper field—a substrate where causes ripple out, consequences align, and reality exists independently of the observer.

That's the leap UED makes.

It moves the anchor back to where it always belonged—not in the frame, but in the field.

UED's Reframe: No Frames, Just Fields

What if the universe never cared where you were standing?

What if time didn't tick differently for different travelers—not because clocks were wrong, but because clocks were never fundamental to begin with?

This is the shift Universal Energy Dynamics (UED) demands. It invites us to abandon the observer-centric maze of frames and coordinates and step into the deeper terrain of fields and ripples.

At the heart of UED lies a simple but transformative idea:

The universe does not operate from frames—it operates from fields.

And that field is not spacetime—it is Structured Vacuum Energy (SVE).

SVE: The True Reference

In UED, the vacuum is not empty. It's not a background. It's the stage, the script, and the director all at once.

Structured Vacuum Energy is a dynamic, ripple-capable medium that exists beneath all motion, matter, and perception. It carries tension, alignment, resistance, and resolution.

It doesn't change depending on who's watching. It doesn't contort itself to satisfy frames. It holds the structure of consequence itself.

This means:

- Motion is not about coordinates changing.
- Speed is not about distance over time.
- Reality is not observer-defined.

Instead, motion is understood as a ripple event—a localized disturbance in SVE, traveling not because of a velocity vector in some inertial frame, but because of energy attempting to restore equilibrium.

What Really Moves?

From the UED perspective:

- An object doesn't "move through space"—its ripple alignment shifts within the SVE field.
- A clock doesn't slow because of speed—it reflects a delay in ripple resonance under field tension.
- A reference frame is irrelevant—because the field is the only consistent reference.

You don't need to compare one observer to another to explain what happens. You only need to know what the field is doing at that location, and how the ripple within it behaves.

Just as a musician doesn't care where in the room the listener is sitting—the notes still resonate in the same key—the universe doesn't change its laws for each observer. It ripples as it always has.

UED reclaims reality from relative frames. It gives it back to the field—the only place where cause and consequence remain intact, no matter who's watching.

Motion as Ripple Alignment

What is motion?

In classical physics, it's defined as change in position over time—a coordinate shift in space. In relativity, it becomes more complicated—motion through curved spacetime, described relative to different observers, distorted by speed and gravity.

But in Universal Energy Dynamics (UED), motion is neither displacement through emptiness nor a trick of the clock. It is something far more elegant and fundamental:

Motion is ripple alignment in flux.

Ripple = Energy in Action Within a Field

In UED, nothing exists in isolation. All entities—particles, planets, photons, you—are ripple patterns within Structured Vacuum Energy (SVE). When something "moves," what's actually happening is a change in the local alignment of ripples with respect to the field.

Think of it like tuning forks in water. When one vibrates, it sends out ripples. If another is tuned to resonate, it begins to hum in harmony. If the field between them shifts, the resonance either intensifies or fades.

This is motion in UED: the energetic shift of ripple interactions through a structured medium, not the mechanical transport of an object from point A to point B.

Displacement Is a Misreading

In traditional views, we talk about velocity—how fast something moves across space. But space, in UED, isn't a container. It's not a backdrop. It is the expression of field structure—and what we perceive as distance is really just a difference in field alignment over a gradient.

So movement isn't about "traveling." It's about realigning with a new field tension configuration.

You don't push through space—you reorganize your ripple signature within it.

The Observer's Motion Is a Shift in Resonance

Under relativity, the observer's motion defines reality. But in UED, the observer's motion does not change the laws of physics—it only changes where and how they resonate with the ripple field.

Two observers might interpret ripple events differently—not because nature changed, but because their point of resonance differs. They're not in separate realities. They're in different zones of alignment.

No Need for Lorentz Transformations

Special relativity uses Lorentz transformations to reconcile time, length, and simultaneity between frames. It requires complex equations to preserve the "speed of light" as constant for all observers, even if their motion differs.

UED needs no such acrobatics.

- There is no need to stretch or shrink space.
- No clocks to dilate.
- No rulers to contract.

Because there is no contradiction in ripple logic. All effects—motion, perception, interaction—are direct consequences of how energy aligns within the SVE field at any given point. The laws don't shift. The observer's alignment does.

It's simpler. Cleaner. Consequential—not conditional.

Motion is not a trick of time or space. It is the dance of energy seeking balance through the ripple.

Consequence Over Perspective

Modern physics often treats perception as a kind of authority. In relativity, what you see—or rather, what your instruments record in your frame—is not just *an* interpretation, but often the defining reality for you. Two observers may disagree on sequence, duration, or even causality, and the theory doesn't say one is right—it says both are.

It's a clever compromise. But it leads to a dangerous implication:

Reality becomes fragmented into points of view.

UED doesn't buy into that. Because truth is not what appears—it's what ripples.

Perception Is Delayed Response, Not Creation

UED makes a bold shift:

It doesn't say, "What you observe is what happens." It says,

"What happens ripples through the field—and whether you notice or not, it leaves consequences."

Events in the universe are not waiting for an observer to validate them. They unfold through field tension, ripple emission, and realignment—with or without eyes watching.

Perception, in this view, is not a cosmic judge. It is simply a localized response to ripple alignment—a sensory echo of a consequence already underway.

The Field Determines Truth, Not the Frame

A star explodes in a distant galaxy. Its light reaches Earth millions of years later. Under relativity, that delay is a function of distance and the constancy of light speed in your frame.

Under UED, that delay isn't a transport time—it's a consequence of ripple realignment. You don't "see" the event—you receive its ripple once your field is ready to engage.

Whether you detect it or not, the explosion happened—it triggered real, irreversible changes in the surrounding ripple field. Those consequences ripple outward, independently of any observer.

You may witness a ripple wave later, but the universe never paused to wait for your eyes.

Perception = Product of Field Tension

In relativity, the observer defines the terms. In UED, the field defines the terms, and the observer aligns or misaligns accordingly.

What you perceive is not the source of truth. It is a friction point—where your local ripple pattern interacts with a larger wave of consequence.

Sometimes you're tuned in. Sometimes you're out of phase. But what's happening doesn't care about your point of view.

UED returns reality to its rightful source: not the eye, not the frame, but the field.

Truth doesn't need to be seen to be real. And physics doesn't need to be framed to be faithful.

Events Without Observers

What happens when no one is watching?

In classical physics, the answer is simple: reality unfolds whether or not anyone is there to observe it. But in modern physics—especially under relativity and quantum interpretations—this simplicity begins to blur. Observation creeps into the equation, and suddenly the act of measurement takes center stage.

But Universal Energy Dynamics (UED) restores clarity with a profound and sobering claim:

Events do not require observers. They require only consequence.

The Event = Field Disturbance

In UED, an event is not defined by perception or frame—it is defined as a disturbance in the Structured Vacuum Energy (SVE).

- A collision.
- A pulse.
- A ripple emitted from imbalance.

These are real, whether or not a witness ever appears. They exist as energetic signatures—patterned deformations within the field.

The field holds the memory. The field carries the consequence. The field is the witness.

Observation Happens After Alignment—Not Before

An observer doesn't make an event real. They only become aware of it after ripple alignment is achieved.

This is where UED introduces the key concept of ATT: Alignment Threshold Time.

- You perceive an event only when your local ripple configuration matches the incoming disturbance.
- If you're not aligned—you miss it.
- If you are aligned—you receive it, not as a transported

image, but as a real-time consequence arriving via the field.

In this framework:

- There's no delay because of distance, only due to alignment lag.
- There's no observer effect—just an observer's position within the ripple field.

You are not causing the event by noticing it. You are simply tuned in to a moment the field has already registered.

No Frames, No Clocks, No Coordinates Needed

Relativity requires reference frames to determine what happened. It needs synchronized clocks, agreed-upon coordinates, and careful calibration to declare who saw what and when.

UED needs none of that.

Because ripple interaction is not dependent on observation—it's driven by field logic. An event's existence doesn't hinge on whether it's measurable. It hinges on whether it rippled the field.

If it did, it's real. If it didn't, it's irrelevant.

The universe does not need to be watched to work. It ripples with or without you.

And when you're finally aligned, you don't see the past. You feel the field—as it is, and as it was always becoming.

The Simplicity of Ripple Logic

Relativity is a masterclass in precision—but it's not simple.

Its beauty lies in its equations, yes—but that beauty often comes at the cost of intuitive understanding. To reconcile different observers, you need Lorentz transformations, tensor calculus, coordinate shifts, and time warps. You end up building a universe of bending clocks and shrinking rulers, just to preserve the illusion of symmetry across frames.

But what if the universe never needed this mathematical gymnastics? What if all this complexity was not a window into reality, but a workaround for misunderstanding its medium?

Universal Energy Dynamics (UED) says:

You don't need to bend time or space. You only need to follow the ripple.

Ripple Interactions Replace Relativistic Transforms

In UED, reality isn't constructed by adjusting one observer's frame to match another's. Instead, it's built from ripple events—interactions of energy in the Structured Vacuum Energy (SVE) field.

Each ripple carries:

- Its intent (a direction of influence),
- Its tension (energy content), and
- Its consequence (how it modifies the field).

What one observer perceives isn't the product of speed or geometry—it's the result of when and how their local field aligns with these ripples.

This is captured through ATT—Alignment Threshold Time:

- You experience something not because it traveled to you through a vacuum,
- But because your field reached the threshold of readiness to engage with it.

No more synchronizing clocks. No more rotating frames. Just ripple logic operating in a universal medium.

No Paradoxes—Just Pattern

One of the most famous outcomes of relativity is the twin paradox:

- One twin travels near light speed, returns younger than the other.

- Time has dilated. Clocks disagree.

In UED, such paradoxes dissolve. There is no contradiction—only a difference in ripple alignment timelines.

Each twin's field experienced different ripple interactions and field tension. Aging, perception, and energetic flow all depend on alignment—not on clock rate or path length.

Similarly, other seeming oddities—length contraction, simultaneity relativity, observer-dependent event order—become non-issues in UED. They were artifacts of how we forced geometry onto ripple-based reality.

Alignment Is the Only Variable

In UED:

- Motion = alignment change in SVE.
- Force = ripple pressure resisting or restoring balance.
- Perception = local resonance with external ripple events.
- Time = the duration it takes for a system to realign after disturbance.

This turns physics from a complex exercise in coordinate matching into a fluid understanding of consequence.

It doesn't simplify because it's lazy. It simplifies because truth doesn't need translation—it just needs clarity.

When you stop chasing frames, you find the field. When you follow the field, you don't need paradoxes to explain reality. Just ripples—and the harmony they seek.

Closing Reflection

"Observers didn't create the universe. But the universe created observers—by rippling just right."

For too long, physics has placed the observer at the center of the story.

It started with good intentions—after all, measurement is essential. But somewhere along the way, we began mistaking the lens for the landscape, the observer for the origin.

Relativity didn't just describe how things look from different vantage points—it made those vantage points the foundation of physical law. It told us that what happens depends on·who's watching, how fast they're moving, and what clock they carry.

But this view is incomplete, and at times, misleading.

Because the universe wasn't waiting to be seen. It wasn't fragmented into frames. It wasn't conditional upon coordinates.

It was already rippling—long before any observer appeared.

UED brings us back to this pre-observer universe—a reality governed not by shifting perspectives but by structured interaction.

You don't need to frame reality to understand it. You need to listen to the field—to the structured vacuum energy that ripples consequences into being, not because someone's looking, but because balance must be restored.

In this view:

- Perception is a result, not a requirement.
- Motion is alignment, not displacement.
- Time is tension resolving—not a universal tick.

And you—the observer—aren't the creator of meaning. You are the emergent harmony of ripple patterns, tuned just right by the universe to notice itself.

Not the cause. Not the frame. But the product of a deeper field that needed no witness to begin with.

The observer didn't shape the cosmos. The cosmos shaped the observer—through ripples, tension, and timeless intelligence.

Chapter 12: The Real Theory of Everything

How force, time, space, motion, and perception arise from a single dynamic

"The universe doesn't run on parts—it runs on patterns."

For over a century, scientists have searched for a final theory—a master equation to unify all forces, all phenomena, all that is.

They called it the Theory of Everything. They imagined it as a fusion of gravity and quantum mechanics. A framework where every particle, force, and field would be explained with absolute precision.

But what if that dream was chasing shadows?

What if the reason we never found the theory of everything is because we started with everything—forces, fields, frames—and never asked if they were real in the first place?

What if the answer isn't in merging what we know, but in unlearning what we've assumed?

Universal Energy Dynamics doesn't propose yet another layer of complexity. It offers a single, foundational shift:

Everything we observe is a ripple event—energy in action—within a structured, intelligent field.

There are no fundamental "things." There are only consequences rippling through a medium that was never empty.

In this chapter, we stop dissecting reality into categories like time, space, motion, and force—and instead trace them all back to one core dynamic:

Ripple behavior in Structured Vacuum Energy (SVE).

No more artificial separations. No more invented entities. No more paradoxes.

Just one field. One rhythm. One reason everything exists—and everything acts.

Beyond the Myth of Unification

Since the early 20th century, modern physics has been in pursuit of a legendary goal: The Theory of Everything.

It's a noble quest. The dream? To find a single framework that explains all forces—gravity, electromagnetism, the strong nuclear force, and the weak nuclear force—as different aspects of the same underlying reality.

Einstein spent the latter part of his life chasing it. So did many giants after him. From string theory to quantum gravity, countless models have tried to unify the universe by merging the forces.

But in doing so, physicists may have been asking the wrong question all along.

The Particle Puzzle

Most unification attempts begin with particles and forces:

- Electromagnetism was unified beautifully by Maxwell.
- The weak force and electromagnetism were combined into the electroweak theory.
- The strong nuclear force joined the standard model.
- But gravity—the most ancient force—refuses to merge.

So the challenge became: How do you make gravity "fit" into the quantum box? How do you make it act like a force when it seems to be more like a curvature, a consequence, a shape?

Instead of asking, "What causes all forces?" We kept asking, "How do we blend their behaviors?"

It was like trying to unify the ripples on the surface of a lake without acknowledging the water beneath.

The Real Oversight: Ignoring the Medium

In all these models, the medium—the actual field in which energy acts—was either oversimplified or ignored.

Space was treated as a stage. Vacuum as empty. Interaction as probabilistic events between point particles.

But what makes interaction possible at all? What sustains the field, carries the forces, holds the energy, transmits the wave?

That's where physics faltered.

It tried to unify the tools while ignoring the table they sit on. It studied the players and missed the orchestra.

UED's Stance: Don't Unify Forces—Understand the Field

Universal Energy Dynamics begins with a single foundational shift:

Everything arises from ripple dynamics in Structured Vacuum Energy (SVE).

You don't need to force gravity and electromagnetism into the same mold. You don't need a common particle, or a fifth force, or a higher-dimensional string. You need to understand how the ripple behaves in different field tensions.

Because what we call different forces are not different ingredients. They are different outcomes of ripple alignment, resistance, and resonance.

The field is already unified. It doesn't need help. It only needs to be heard—without translation, without reduction.

The universe isn't fragmented. Our theories were.

The One Dynamic: Ripple in Structured Vacuum Energy (SVE)

Let's remove all the labels—particles, waves, forces, fields, even space and time—and ask: What is truly happening in this universe?

At its core, there is only one phenomenon that is always present when anything happens: A ripple.

A ripple is energy in motion. But not motion through space—as traditionally imagined. Rather, it is energy disturbing, stretching, and seeking re-alignment within the field itself. And that field is not space, nor some abstract background.

It is Structured Vacuum Energy (SVE)—the invisible, intelligent medium that holds the universe in coherence.

The Three States of Reality

In UED, all of reality can be described by the state of SVE:

1. Stable – Perfect balance. No motion. No time. No perception. → This is the ground state—the resting pulse of the cosmos.
2. Disturbed – Ripple is introduced. Energy becomes active. → This is the start of consequence—where experience, change, and form begin.
3. Re-aligning – The field self-adjusts, seeking balance. → This is what we perceive as evolution, decay, motion, time, and force.

Every atom, every breath, every galaxy is a temporary expression of these states.

Ripple = The Only Real Dynamic

There is no "thing" moving through "space." There is only ripple activity—energy attempting to restore equilibrium through structured tension.

Ripples are:

- Not waves in a medium—they are the medium becoming dynamic.
- Not particles—they are point-like concentrations of ripple resonance.
- Not waves *and* particles—they are the cause behind both illusions.

The ripple is not a metaphor. It is the foundational behavior of energy.

It is how the field reacts to imbalance, how it communicates, how it manifests everything else.

Emergence: How Everything Arises from Ripple Behavior

From this one dynamic, the entire orchestra of reality unfolds:

- Particles are ripple knots—stabilized resonances in tension zones.
- Forces are not carriers—they are ripple gradients.
- Fields are structured zones of ripple potential within SVE.
- Space is the distance between ripple nodes in tension.
- Time is the lag it takes for a ripple to align with or return to equilibrium.

There is no need to invent multiple rules for each phenomenon. Because there aren't multiple phenomena—only multiple expressions of one field, one ripple, one rhythm.

We kept looking for many laws. But the universe has always been whispering just one— "Disturbance must resolve."

Force = Ripple Tension

In traditional physics, force is treated as a fundamental building block of reality. Newton defined it as a push or pull. Einstein reinterpreted gravity as curvature. Quantum mechanics speaks of force carriers like photons, gluons, and the yet-to-be-seen gravitons.

But in all these views, force remains an external agent—something acting *upon* objects, making them move, fall, or accelerate.

Universal Energy Dynamics (UED) flips this notion inside out.

Force isn't something acting from outside. It is the internal resistance of a ripple trying to restore harmony within the field.

Force Is Not a Thing—It's a Tension

In UED, everything begins with Structured Vacuum Energy (SVE). When the field is perfectly aligned, no force is felt—there is stillness, balance, silence. But when there is a disturbance, a ripple forms.
That ripple carries tension—a desire to return to equilibrium. What we call force is not the cause of motion—it is the resistance to misalignment, the pressure the ripple exerts as it seeks balance.
Think of it like this:

- You pull a spring—it resists.
- You stretch a rubber band—it tightens.
- But in UED, this isn't happening to objects. It's happening to the field itself.

Wherever there is ripple misalignment, there is force tension.
Gravity = Gradient of Ripple Equilibrium
We once imagined gravity as a force pulling objects together. Then Einstein taught us to see it as the bending of spacetime. But UED sees gravity as even more fundamental:
Gravity is not a pull. It is a slope in the ripple field—an energetic gradient caused by imbalance.
A mass creates ripple tension in the SVE. Another mass near it naturally moves along this gradient—not because it's attracted, but because it is seeking alignment within the disturbed field. No graviton needed. No mystery.
The more intense the imbalance, the steeper the ripple slope. That's why massive bodies curve reality—not geometrically, but energetically.
Electromagnetism = Directional Ripple Flow
Electric and magnetic fields aren't separate in UED—they are orientations of ripple flow.

- Charge arises from asymmetry in ripple alignment.
- Polarity is a matter of direction in ripple propagation.

- Attraction and repulsion emerge from how ripple fronts interfere, amplify, or cancel.

An electric current is simply a coordinated ripple channel. Magnetism is spiral ripple structure resulting from that flow.
You don't need virtual particles to explain interaction. You just need ripple logic and a field that knows how to carry tension across space.

The Illusion of Force Carriers

Modern physics leans on force carrier particles:

- Photons for electromagnetism
- W/Z bosons for the weak force
- Gluons for the strong force
- A theoretical graviton for gravity

These are not wrong—they are mathematical conveniences that help preserve conservation laws and explain energy transfer.
But UED asks:
What if there are no messengers—only medium?
What if all forces are just expressions of ripple tension—no particles flying across space, no exchange diagrams, just real-time ripple interaction within a responsive, intelligent field?
Then what we called "forces" were never things. They were experiences of imbalance—and the universe's persistent attempt to resolve them.
Force doesn't push you. The field simply refuses to stay misaligned.

Space = Ripple Configuration

We've been taught to think of space as a vast, three-dimensional container—an empty arena where matter floats, forces act, and time ticks. In both Newtonian and Einsteinian models, space is the stage: static or curved, but fundamentally separate from the events that occur within it.

But what if this idea of space as a passive backdrop is fundamentally flawed?

Universal Energy Dynamics (UED) offers a radical reinterpretation: Space is not a container. It is a configuration of ripples in Structured Vacuum Energy (SVE).

There Is No "Empty" Space—Only Structured Field

In UED, the universe is not floating in emptiness. There is no void. No blankness to fill.

Instead, every "where" is simply a pattern in the ripple field. Even what appears to be empty is alive with ripple stability zones, tension gradients, and field symmetries.

You don't exist in space. You are part of the field configuration.

When you move, it's not through a gap— It's a re-alignment of your ripple signature within the surrounding topology.

Distances Are Field Tensions, Not Fixed Lengths

In conventional physics, distance is a number—a straight line from point A to point B. But in UED, there is no such thing as absolute distance.

Instead:

- What we call "distance" is a measure of ripple complexity between two sources.
- It reflects the energy required for alignment or interaction to occur between two ripple zones.
- If the field is smooth between two points, the perceived distance is short. If it is turbulent or distorted, the "distance" stretches—not because the metric changes, but because alignment becomes harder.

So distance is not a number—it's a function of field readiness.

This explains:

- Why gravity affects perception of space.

- Why fields can be "stretched" or "compressed."
- Why space can "expand" without new matter appearing.

Space Is Ripple Stability, Not Separation

We often imagine space as what separates objects. But in UED, space is what holds their relationships. It is the pattern of stability zones around ripple sources.

Each particle, planet, or photon generates a ripple field. Where those ripples settle into equilibrium, we experience spatial structure.

What we perceive as "volume," "position," or "geometry" is simply a map of ripple stabilization.

The more complex the pattern, the more dimensional the space appears. But it's not about dimensions—it's about energy resonance boundaries.

Space is not the absence of things. It is the presence of patterned stillness in a sea of active possibility.

Space is not where things are. It's how the field has chosen to hold them apart—without breaking the rhythm.

Time = Delay in Realignment (ATT)

What is time?

It's the most familiar mystery we live with. We schedule it, measure it, chase it, waste it—yet we hardly understand it. Physics treats it as a dimension. Clocks treat it as uniform ticks. Relativity bends it. Quantum theory barely knows what to do with it.

But Universal Energy Dynamics (UED) invites a profound shift:

Time is not something that flows. It is something you experience—because the field hasn't fully aligned yet.

Time Is Not Universal—It's a Consequence

In Newton's world, time was absolute: same everywhere, for everyone.

In Einstein's world, time was relative: it stretched and shrank depending on speed and gravity.

But both assumed one thing: that time exists as a backdrop—either flat or curved.

UED challenges that assumption at its root.

Time is not a background. It is not a substance. It is not ticking anywhere "out there."

Instead, time is the delay between disturbance and realignment in Structured Vacuum Energy (SVE). It is how long it takes for the ripple to settle. It is the temporal echo of imbalance.

ATT — Alignment Threshold Time

UED introduces ATT: Alignment Threshold Time, the smallest meaningful measure of experienced change.

ATT is not based on clocks or frequencies—it is based on field readiness:

- When a ripple disturbs the field, there is a delay before it aligns with your local structure.
- You experience the event only when your field reaches the threshold to engage that ripple.
- Different systems have different ATT values based on ripple complexity, density, and local field tension.

This means:

- You don't perceive time—you perceive the lag in ripple resonance.
- The more misaligned the field, the longer the ATT—which we interpret as "slower time."
- In perfect alignment (deep presence, flow states, or enlightened awareness), ATT collapses—and time "disappears."

No Need for Clocks or Spacetime Curves

In relativity, time slows near gravity wells or at high velocities. But what is gravity, really? In UED, gravity is field misalignment—and so is the resulting time distortion.

So you don't need curved spacetime to explain time dilation. You only need to understand how ripple delay increases with field tension.

Likewise, you don't need a cosmic clock ticking uniformly for all. Each ripple system experiences time based on its own ATT inertia—how quickly or slowly it can adapt to field changes.

This explains:

- Why time "feels slow" in pain or trauma (high ripple turbulence).
- Why time "flies" in joy or immersion (low field resistance).
- Why time varies not just in physics—but in psychology, biology, and emotion.

Because time is not a line—it's a lag. A lag between intent and alignment. Between disturbance and peace.

You don't move through time. Time moves through you—because your field hasn't finished responding yet.

Motion = Shift in Ripple Alignment

We grow up thinking motion is simple: a ball rolls, a bird flies, a car speeds up—something moves from one place to another across space.

Physics formalized this with velocity, acceleration, and frames of reference. Newton gave us equations. Einstein made us rethink speed limits. Quantum theory made it fuzzy.

But Universal Energy Dynamics (UED) offers a deeper clarity:

Motion is not movement through space. It is a shift in ripple configuration within Structured Vacuum Energy (SVE).

There Is No "Travel"—Only Realignment

In conventional physics:

- An object moves from point A to B.
- We define its position using coordinates.
- We calculate its velocity based on time and distance.

But UED says: there is no absolute "A" or "B." There is only the pattern of ripple alignment at one moment, and the pattern it transitions into the next.

Motion = the field's reconfiguration of tension.

It's not about something being "carried" across emptiness. It's about the field restructuring itself so the ripple expression appears in a new place.

Just as an image on a screen isn't physically moving—the pixels are just switching— motion in UED is the universe shifting its ripple expression without needing anything to "travel."

No Need for Velocity Vectors or Frames

In relativity, motion is defined relative to a frame—your speed depends on who's watching, and clocks may tick differently.

But if space is ripple configuration and time is ATT, then motion isn't measured—it's felt as change in ripple balance.

There is no need to:

- Track distances through geometry.
- Compare speeds between observers.
- Worry about reference frames.

Instead, we ask:

- Has the ripple pattern shifted?
- Has the field realigned to a new tension state?

If yes, motion has occurred—even if no object "traveled."

Motion Emerges from New Alignment Zones

Wherever the ripple field settles into a new zone of stable tension, we perceive it as "arrival."

Wherever it breaks free from an old zone, we perceive it as "departure."

Motion is not about momentum—it is about transition. It is not a push or pull—it is the unfolding of field readiness to accept a new configuration.

This explains:

- Why electrons "jump" between states.
- Why light can be everywhere at once—and yet not arrive until alignment.
- Why consciousness can move without physical displacement (e.g., memory recall, imagination, intuition).

Because the true mover is the field, not the thing.

Motion isn't the path between places. It's the field deciding to ripple differently.

Perception = Resonance with Ripple Events

What is perception?

We usually describe it as the act of sensing something: seeing light, hearing sound, touching texture, feeling time pass. It seems like the world flows into us—through eyes, ears, skin, and thought.

But in Universal Energy Dynamics (UED), this idea is reversed:

You don't receive the world—you resonate with it. Perception is not passive reception—it is active alignment.

The world is not coming *at* you. You are tuning *into* the world—through ripple behavior.

You Perceive What Your Field Can Align With

Everything you perceive—an image, a song, a thought—is a ripple event in Structured Vacuum Energy (SVE). But not all ripple events reach you. And not all that reach you are perceived.
Why?
Because your local field—your body, nervous system, consciousness—is a complex ripple structure. It can only engage with ripple events that match its current state of alignment.
This is resonance.

- If your field aligns with a ripple pattern → you experience it.
- If it doesn't align → you miss it completely.

This explains:

- Why you notice things only when you're "in the right mindset."
- Why two people can experience the same event differently—or not at all.
- Why deep perception feels like "tuning in," not absorbing.

Perception is the meeting point of internal and external ripples. No alignment = no awareness.
Light, Sound, Experience = Ripple Resonance
In classical science, light is a wave or a particle sent from a source to your eye. Sound is air pressure oscillation reaching your ear. Experience is neurons firing.
In UED, these are all ripple interactions:

- Light is the ripple realignment of SVE triggered by energetic resonance.
- Sound is a structured ripple pattern in dense field zones (like air or matter).
- Experience is the field's internal ripple pattern aligning with

an external event.

You don't "see" light because a photon hit your retina. You see because your field reached the ATT where its alignment matched the ripple that was already present.

Everything you sense is the product of a tuning event—a harmonic interaction between field structures.

There is no delivery of reality. There is only real-time ripple agreement.

Consciousness = Tuning Capacity of Ripple Complexity

So what is consciousness?

UED proposes a fresh definition:

Consciousness is the capacity of a ripple system to tune into increasingly complex ripple patterns.

It's not magic. It's not a byproduct. It is a layered ripple phenomenon—a dynamic, self-aware field capable of aligning with other ripple events across many scales.

The more intricate the ripple structure (like the human brain), the more subtle and abstract the resonances it can engage with:

- Emotions
- Memories
- Ideas
- Ethics
- The sense of time and self

This makes consciousness a refined ripple lens, capable of perceiving reality not just as it appears, but as it unfolds across dimensions of resonance.

That's why states of consciousness—sleep, dream, alertness, flow—shift what we perceive. They re-tune our ripple capacity.

You don't perceive because something happened. You perceive because your field became ready to resonate with what was always there.

Why This *Is* the Theory of Everything

For decades, scientists have searched for a "Theory of Everything"—a single formula, a grand equation, that could tie together gravity, quantum mechanics, thermodynamics, cosmology, and consciousness into one clean, coherent package.

The more we searched, the more complex it became:

- More dimensions.
- More particles.
- More exotic forces.
- More paradoxes needing patches.

But what if the answer wasn't about adding more layers?

What if the real Theory of Everything begins by removing the illusions?

The Mistake: Looking for Unity in Parts

Mainstream physics tries to unify by combining components:

- Merge all the known forces.
- Explain particles through strings.
- Fit dark matter into gaps.
- Adjust equations to match observations.

But in doing so, it assumes the parts are real to begin with.

UED challenges this foundational assumption. It doesn't unify forces—it dissolves the illusion that they were ever separate. It doesn't patch over contradictions—it shows they arise because we misunderstood the medium itself.

UED's Power: Subtract Illusions, Reveal the Field

Universal Energy Dynamics proposes a radical elegance:

All observed phenomena—force, space, time, motion, perception—are ripple consequences within a single, structured field.

You don't need:

- A graviton to explain gravity.
- A multiverse to explain probabilities.
- Dark matter to explain galaxy rotation.
- Imaginary time to explain the Big Bang.
- Quantum magic to explain consciousness.

Everything becomes visible, understandable, and testable when you see it as a ripple tension resolving through structured vacuum energy (SVE).

UED doesn't create new things—it reinterprets existing ones through consequence-based logic.

It doesn't add—it subtracts what isn't real.

From Equations to Understanding

Science has often pursued a master equation—some symbolic expression that contains the secret of reality.

But UED says:

The Theory of Everything isn't a formula. It's a framework of understanding.

A ripple doesn't need coordinates to act. It doesn't need a particle zoo to behave. It doesn't care about frames, reference points, or clocks.

It follows only one rule:

- Disturbance must resolve.
- Misalignment must realign.
- Ripple must echo, adjust, and return to balance.

From this rule, everything flows: light, time, gravity, thought, birth, death, memory, matter.

The universe wasn't built from parts—it unfolded from patterns. And the ripple is the pattern beneath all others.

This isn't just physics. It's a return to truth without contradiction.

Closing Reflection

"In the beginning, there was no matter, no motion, no time. There was only the field. And the first ripple—was everything."

Before the stars were born, before light flashed through cosmic silence, before time began ticking—there was the field.

Not empty. Not passive. But whole. Poised. Structured.

It needed no dimension to exist, no observer to witness it, no particle to define it. It simply was—a perfectly balanced presence, like a still lake before the first drop falls.

And then... it rippled.

Not from intention. Not from accident. But from the intrinsic tension of being capable of change.

That first ripple was not a thing—it was an event. A difference. A deviation. A movement without motion.

From that single act of imbalance—time began. From the consequence of disturbance—space emerged. From the dance of returning balance—forces took shape. And from structured resonance—consciousness would one day open its eyes.

Everything we call physics, chemistry, biology, or mind... All of it is just a trail—ripples echoing from that first disturbance, like aftershocks that never stopped.

We tried to name the ripples:

- Quarks.
- Photons.
- Gravity.
- Love.

- Light.

But beneath every name was a rhythm. And beneath that rhythm was the field, still holding it all together.

We were never separate from it. We were never outside of it.

We are the field, realizing itself through ripple.

So if you seek a theory of everything, stop counting particles. Stop drawing strings across imagined space. Stop bending time to fit equations.

Instead, return to the only thing that ever moved— The ripple.

It was enough then. It is enough now. It is enough to explain everything.

There was no Big Bang. There was just a perfect stillness... ...and a ripple that changed everything forever.

Part IV: The Reawakening

Chapter 13: The Observer Returns

Mind, consciousness, and the forgotten role of perception in physics

"We never left the experiment. We are the field measuring itself."

Physics began with a paradox it has spent centuries trying to ignore:

You can't study the universe without being part of it.

Yet, in the pursuit of objectivity, science exiled the observer—declaring the mind a contaminant, consciousness an afterthought, and perception a distortion. The universe, we were told, must be understood as if no one is watching.

But the irony is sharp and undeniable:

Every observation, every equation, every theory—was filtered through a mind, shaped by perception, and delivered by awareness.

We never left the experiment. We just pretended we weren't part of it.

In classical physics, the observer was erased to keep the math clean. In quantum mechanics, the observer reappeared like a ghost—collapsing probabilities without ever being explained. In relativity, the observer became a reference frame—disembodied, floating, but still crucial.

And yet, no one asked the deeper question: What is this observer? What is mind? What is perception, really? Why does it matter what is seen—and by whom?

Universal Energy Dynamics (UED) doesn't treat the observer as a nuisance or a side effect. It recognizes the observer as a ripple-sensitive structure within the field itself. Not just a watcher—but a consequence-aware participant. Not outside the system—but the very field remembering itself through resonance.

In this chapter, we don't just bring the observer back. We show that it was always central. That mind, consciousness, and awareness are not accidents of biology—but essential expressions of field intelligence.

Because reality isn't complete until it's aligned with awareness.

And awareness, as we'll see, is nothing but the field reaching back into itself—with eyes wide open.

The Vanishing Observer

At the dawn of modern science, a great sacrifice was made: To make physics "objective," we removed the one thing that could never truly be separated from it—the observer.

This decision shaped the entire arc of scientific development. The observer—the very agent of knowledge—was demoted to a neutral instrument or statistical anomaly, tolerated but never explained.

Let's trace how this exile unfolded.

Classical Physics: Erasing the Witness

In Newtonian mechanics, the universe was a vast, predictable machine. Events happened independently of who was watching. Motion, force, and mass obeyed universal laws, the same for everyone, everywhere.

To achieve this mechanical objectivity, the observer was deliberately removed. The mind was messy. Perception was fallible. Emotion had no place.

So physics chose silence over complexity. It assumed:

"If we can measure it, it's real. If we can't, it's irrelevant."

This helped build bridges, launch rockets, and power industries. But it also created a blind spot: The one doing the measuring—the mind itself—was left outside the equation.

Quantum Mechanics: The Ghost of the Observer Returns

Then came the 20th century—and the neat picture shattered.

In quantum mechanics, the observer could no longer be ignored. Particles did not behave like classical objects. Their states were not determined—until observed.

This birthed the infamous concept of wavefunction collapse:

- Before measurement, a quantum system exists in multiple states (superposition).
- Upon observation, the wavefunction "chooses" a definite outcome.
- The observer, it seemed, triggered reality itself.

Suddenly, the act of looking was an act of creation.

But even here, the observer was treated like a magical switch—not a field-aware system. Physicists called it "measurement," avoided defining consciousness, and left questions like:

Who or what counts as an observer? Does a camera collapse the wavefunction? A cat? A human? A rock?

...completely unanswered.

The observer was present—but still not understood. Still outsourced from the theory.

Relativity: The Frame Game

Einstein shifted the focus from who observes to how they move.

In Special Relativity:

- Time and space changed depending on your velocity.
- Simultaneity vanished.
- There was no absolute frame—only frames of reference.

This decentralization sounded democratic. But again, it turned the observer into a floating coordinate system—not a conscious agent. Your perception of time could stretch or shrink, but no one asked why you perceive at all.

Relativity treated the observer as a mathematical placeholder, not a mind. Just a node with a clock, not a living being embedded in the field.

What Was Missing?

Across all these revolutions—classical, quantum, relativistic—perception itself was never examined.

What is perception? What is mind? Why do some systems perceive while others don't? How does observation emerge from the field? What role does alignment, resonance, or consciousness play?

These questions were either ignored or offloaded to psychology, philosophy, or metaphysics—territories "outside" physics.

But the irony is obvious:

Theories meant to explain reality excluded the very experience of reality itself.

The observer didn't just vanish. It was written out of the script to protect the illusion of objectivity.

Now, with Universal Energy Dynamics, the observer is no longer a glitch. It is the final proof that physics was never complete without the mind.

Perception Is Not a Distortion—It's a Ripple Interface

For too long, science has treated perception as something unreliable. A distortion. A filter between "real" reality and our limited human senses.

We've been told:

- The world is objective.
- Our minds only give us approximations.
- What we perceive is not what truly exists.

This narrative separated truth from experience, implying that reality is "out there," pure and untouched—while perception is subjective, messy, and flawed.

But Universal Energy Dynamics (UED) offers a radical shift:
Perception is not a flaw in the system. It is the system at work.
Perception is not a veil—it is a ripple interface. Not an illusion—but an alignment.

Perception = Field Interaction, Not Passive Reception

In UED, all existence is ripple-based. Every object, every event, every moment is a ripple pattern in Structured Vacuum Energy (SVE).
Perception doesn't happen when light "hits" your eyes. It happens when your internal ripple field aligns with an external ripple event.
You don't just "receive" information. You tune into it.
Your senses are not just detectors. They are field alignment organs, capable of:

- Receiving
- Resonating
- Translating ripple interactions into experience.

This means:

- If there's no ripple match, there's no perception.
- If there's partial match, you get distortion.
- If there's full alignment, you get clarity—not just sensory clarity, but cognitive, emotional, even spiritual.

In UED, this resonant alignment is the true definition of perception.

You Don't See Events—You Resonate with Ripples

Let's say a supernova explodes in a distant galaxy. Did you "see" it? No. A ripple pattern aligned with Earth's SVE field. That pattern, over time, resonated with your retinal field, and your brain's internal ripple map processed that resonance.
From that alignment—an image emerged in your consciousness.
The supernova didn't travel to you. Its ripple signature met your awareness through alignment thresholds (ATT).

This applies to everything:

- Sound: You don't hear vibration in air—you resonate with structured ripple compression.
- Touch: You don't feel atoms repelling—you resonate with resistance in ripple density.
- Emotion: You don't just react—you feel the echo of previous ripple misalignments surfacing.

Perception is not reception—it's participation in the ripple field.

The Observer Is a Node of Alignment—Not Outside Reality

Physics often modeled the observer as separate:

- A detached measurer.
- A floating reference frame.
- A neutral clock-holder.

But UED rewrites this narrative:

The observer is not outside the system. They are a ripple system embedded within it—capable of aligning with other ripple zones.

You are not looking at the universe from the outside. You are the universe, in ripple form, becoming aware of itself.

This makes perception sacred—not mystical, but energetically grounded.

It also redefines "truth":

- Truth is not what exists independent of the observer.
- Truth is what resonates across the field without contradiction—even when multiple observers align differently.

Perception is not a flaw in reality. It is how the field remembers itself, one resonance at a time.

Consciousness as a Ripple Complexity

In mainstream science, consciousness remains the last great mystery—a phenomenon often described as "emergent," but never truly explained.

Neuroscience maps brain regions. Cognitive science tracks behavior. AI simulates responses.

But no theory can yet answer the fundamental question:

Why does awareness arise at all?

Consciousness is often treated as a byproduct of biological complexity—something the brain accidentally produces after enough neurons fire.

But Universal Energy Dynamics (UED) offers a transformative view:

Consciousness is not an emergent property of matter. It is a ripple network achieving internal resonance within the field.

It doesn't emerge from the brain—it expresses itself through ripple harmony. It doesn't need neurons—it needs structure, alignment, and memory within the ripple topology of the field.

The Brain = High-Density Ripple Harmonizer

In UED, the brain is not a generator of consciousness. It is an incredibly dense ripple system, uniquely capable of:

- Holding long-standing ripple patterns.
- Creating internal simulations of external events.
- Comparing ripple memories with live field input.

Each neuron is not a switch—but a node of ripple alignment, able to hold tension and release it in resonance with the structured field.

The brain's role is not to create awareness, but to stabilize ripple interaction with high resolution and continuity.

It is the field's way of building a localized intelligence center, not to observe passively—but to align, learn, and adapt.

Awareness = Synchronization Across Ripple Thresholds (ATT)

So when do you become aware of something?

When internal ripples and external ripple events reach synchronization—specifically through Alignment Threshold Time (ATT).

If the brain's internal ripple field matches the signature of an external ripple disturbance:

- The field synchronizes.
- The ripple stabilizes.
- You "become aware."

This process isn't mystical—it's energetic coherence.

It's why:

- Awareness is slow in sleep (field desynchronized).
- Sharp in emergencies (field hyper-attuned).
- Elevated in meditation (low turbulence, high coherence).
- Fragmented in trauma (ripple loops block synchronization).

Your state of consciousness is how your internal ripple matrix aligns with field complexity—nothing more, nothing less.

Observation = Consequence-Aware, Not Neutral

Here lies one of UED's most powerful revelations:

There is no such thing as a neutral observation. Every act of observation is a ripple consequence.

When your ripple field aligns with another, it doesn't just "see"—it changes. It becomes part of the ripple history. Your awareness shifts field behavior—not because you collapse a wavefunction, but because your field became an active participant in the alignment.

This isn't metaphysical—it's field logic:

- You align → the ripple completes → the field updates → experience happens.

So the observer is not a passive outsider. They are an interference pattern in the unfolding of field dynamics.

You do not observe reality from a distance. You experience consequences from within the same ripple network.

That is why consciousness cannot be ignored by physics any longer—it anchors what is seen, and decides what becomes real to the observer.

Consciousness is the tuning fork of the field. When it hums in harmony, awareness blooms. When it fractures, reality distorts. But in either case—it is the field experiencing itself.

The Mind as a Mirror of Field Tension

What exactly is the mind?

In everyday life, we treat the mind as a stream of thoughts, feelings, memories, and judgments—something both intimate and elusive. Science often reduces it to a neural process, while philosophy debates whether it is a soul, an illusion, or a function of language.

But within the Universal Energy Dynamics (UED) framework, the mind has a far more grounded and elegant definition:

The mind is the localized ripple interface that mirrors the history of field alignment within and around the self.

It is not a byproduct of neurons. It is not a distortion of reality. It is the echo of ripple tension—constantly responding, adapting, and re-patterning itself based on internal and external interactions.

Thoughts, Memory, and Emotion = Ripple Traces

In UED, your thoughts are not abstractions. They are internal ripple formations replaying or simulating alignment scenarios:

- Thoughts are reactivations of potential ripple paths—mental alignment rehearsals.

- Memory is stored ripple tension—stable patterns from past alignments.
- Emotion is the ripple feedback from misalignment or resonance—energetic consequences of field interactions.

Every mental state you experience is a ripple signature built from prior exposure to field events.
This is why:

- Trauma loops repeat as tension patterns not fully resolved.
- Positive emotions uplift because they ease internal ripple tension.
- Complex reasoning requires internal harmonization of multiple ripple threads.

You are not just thinking—you are feeling the memory of the field.

The Mind Is Not a Subjective Distortion—It Is Field Responsiveness

In classical models, the mind is often seen as subjective, thus unreliable. But UED reframes this:
The mind is not a distortion of reality. It is a dynamic map of ripple responsiveness—updated by every alignment and misalignment you've ever experienced.
Your opinions, beliefs, fears, hopes—these aren't delusions. They are residue patterns from ripple tension your system has endured or embraced.
The mind is like a mirror—but not a passive one. It is a living interface that reflects the tensions of your internal and external environment.

Perceptual Biases = Misaligned Ripple Habits

So what causes bias, misjudgment, or illusion?
From a UED perspective, these aren't "failures" of brain mechanics. They are habitual ripple misalignments:

- When you misinterpret someone's words, it's because your field remembers a ripple pattern from the past.
- When you're closed to new ideas, it's because your field defends old stabilizations.
- When you generalize or stereotype, it's your mind trying to pre-align based on incomplete ripple resonance.

Bias is not ignorance—it is frozen alignment, formed under past field pressure, now resistant to change.

That's why awareness alone can break a bias—not because it adds information, but because it reopens alignment capacity.

True Intelligence = Ripple Mastery

And now comes a deeper insight into what intelligence really is.

Not IQ. Not memorization. Not logical speed.

But this:

True intelligence is the ability to align your ripple field appropriately—with precision, flexibility, and awareness—across varying field conditions.

It's knowing:

- When to engage, and when to rest.
- When to react, and when to reflect.
- When to stabilize, and when to disrupt.

Intelligence is ripple agility. Wisdom is ripple discipline. Conscious growth is progressive mastery over alignment.

The most intelligent minds are not just analytical—they are tension-aware, field-sensitive, and alignment-ready.

Your mind is not a container for thought. It is a field memory organ—rippling with the weight of every alignment you've ever survived, shaped by every tension you've ever resolved.

Physics Without Perception Is Incomplete

For centuries, physics has strived for objectivity—models that could stand independent of human experience. The goal was noble: to describe nature in terms that were universal, measurable, and replicable, regardless of who observed them.

But in its pursuit of detached clarity, physics made a critical oversight:

It tried to erase the observer—while relying on observation to validate reality.

This contradiction has haunted science from the quantum scale to the cosmological. And it becomes glaringly obvious once we examine the role of perception, awareness, and alignment in the very act of doing physics.

Measurement Without Awareness = Blind Models

Every experiment is, at its heart, a conversation between an instrument and an observer. But conventional physics assumes that once we "measure" something, the job is done—awareness is irrelevant.

This approach gives us formulas, data, predictions. But it also gives us blind spots.

Why? Because a measurement without awareness misses the field state of the one measuring. It ignores the ripple conditions necessary for the phenomenon to become visible in the first place.

- You don't just "record" events.
- You realign with them through perception.

So when physics tries to isolate reality from the observer, it ends up isolating reality from reality—splitting truth into fragments that never quite resolve.

Paradoxical Collapse: When Observers Are Ignored

Quantum mechanics was the first to crack open the illusion of observer-independence.

The famous Schrödinger's cat thought experiment highlighted this dilemma:

- A system exists in a superposition of states.
- The act of observation collapses the superposition into one outcome.
- But if no one observes... what happens?

Later refinements like Wigner's friend took it further:

- Two observers can witness different outcomes.
- Who is "right"? Which version is real?

These are not just philosophical puzzles. They are proof that observer-free physics leads to logical paradox.

And why? Because the models ignore ripple alignment—the fundamental basis for how phenomena become perceivable.

In other words:

It's not about who looked. It's about who aligned.

UED: The Observer as a Carrier of Ripple Resonance

Universal Energy Dynamics doesn't fall into this trap.

It doesn't reduce the observer to a button that collapses waves. Nor does it turn them into a ghost in the machine.

Instead, UED defines the observer as:

A carrier of ripple resonance—a localized field structure capable of aligning with other field events.

The observer is not the cause of reality. They are a participant in reality's reveal, a node through which alignment becomes perception.

This view dissolves paradoxes:

- No need to collapse probability clouds—only align with ripple states.

- No confusion over different observers—each perceives based on unique alignment, without contradiction.
- No magic in measurement—just field interaction made visible through synchronized resonance.

Alignment = The Gateway to Visibility

Here's the ultimate truth UED brings to the table:

Nothing is visible unless aligned. Nothing is perceived unless ATT is met. Nothing is "measured" until resonance occurs.

In this framework:

- A particle isn't "there" until your field aligns with its ripple tension.
- A force doesn't "act" until there's a measurable misalignment.
- Time doesn't "pass" until your field completes its ripple delay.

This doesn't mean reality is subjective. It means reality is revealed through consequence-aware alignment.

You don't make reality by observing it. You encounter what's already there—through your ability to resonate with it.

Thus, physics without perception is incomplete not because perception adds magic, but because perception reveals the structure that was always rippling underneath.

We didn't invent reality by watching it. But we can't understand it without aligning with what the field is ready to show.

The Field Knows Itself Through Us

What if we were never meant to be outside observers? What if the very act of observing wasn't a disruption—but a continuation?

Science often imagines us as minds trapped inside biological machines, peering out into a foreign universe, trying to decode its language. But Universal Energy Dynamics (UED) reframes this entirely:

We are not separate from the field. We are how the field becomes aware of itself.

Our minds, bodies, and experiences are not accidental byproducts of biology. They are ripple expressions, shaped by Structured Vacuum Energy (SVE) to serve a deeper function: Self-recognition.

We Are Expressions of Field Intelligence

Every ripple that flows through us is not just energy—it is intelligence in motion. Our nervous system is a precision-tuned ripple interface, evolved not to survive randomly but to sense alignment, adapt to tension, and seek coherence.

This means:

- Your consciousness is not *in* your brain.
- Your brain is a ripple-density hub *within* the field.
- And you—as a conscious being—are a ripple node carrying intelligent alignment history.

You are not watching the universe. You are the universe watching itself from the inside.

This is not poetic metaphor. It is literal in the UED framework. Your thoughts, emotions, and awareness are not noise in the equation. They are how the field updates itself in real time.

The Observer Is the Field Remembering Itself

What is observation, if not memory regained through alignment?

Each time your mind perceives, it doesn't invent. It recalls a ripple structure it is capable of resonating with. This is why perception is sometimes instant, and sometimes delayed—because the field is still reaching for coherence.

So the observer isn't a foreign agent, disturbing a pristine system. The observer is the field folding back into its own ripple—a loop of self-reference, awareness, and realignment.

Every act of seeing, knowing, feeling is the field saying to itself: "Yes, I recognize this. I am this."

True Objectivity = Deep Alignment, Not Detachment

For centuries, objectivity meant distance—removing bias, emotion, and self-interest. But this ideal led to the rejection of the most essential tool of truth: perception through alignment.

UED proposes a new definition:

Objectivity is not detachment. It is resonance with the field across layers.

The deeper your alignment:

- The clearer the ripple pattern.
- The less distortion in perception.
- The more universal the consequence you experience.

True objectivity happens not when you step back, but when you step deep enough into the field that your observations reflect its actual structure, not just your fragmented mind.

From Measurement to Ripple Coherence Awareness

Science must evolve.

It cannot remain obsessed with only what can be counted, weighed, or timed. It must learn to perceive alignment, feel ripple memory, and trace coherence patterns that do not show up on instruments—but shape what those instruments measure.

This doesn't mean abandoning rigor. It means expanding the definition of rigor to include:

- Attunement of the observer.
- Impact of field conditions.
- Awareness of ATT and ripple delays.

We are entering a new era of physics—one that doesn't discard consciousness as noise, but sees it as the quiet intelligence of the field made visible.

The universe doesn't need us to explain it. But it does use us to know itself—through ripple coherence, through memory, through alignment.

Closing Reflection

"We searched the stars for truth. But the instrument was always within. Not the telescope, but the mind— Not the lens, but the alignment."

We built telescopes to peer into galaxies. We smashed particles to find the fabric of matter. We built atomic clocks to chase the heartbeat of time. We thought the answers were out there—in distance, in data, in devices.

But we forgot the one tool through which all questions emerged: The observer. The self. The mind aligned with the field.

For every discovery in science, there was not just an experiment. There was a moment of awareness—when a ripple pattern aligned with consciousness, and a phenomenon became visible.

We gave credit to instruments. But it was never just the telescope. It was never just the equation. It was always the alignment—between what existed and what was ready to be perceived.

We thought we were detached analyzers of a passive universe. But we are—and have always been—participants in its unfolding.

Your mind is not a spectator. It is a mirror of the field, shaped by resonance, sculpted by consequence, moved by ripple tension.

And every time you notice, realize, or reflect—you are not watching from the sidelines. You are the field saying:

"This too, I am." "This too, I feel." "This too, I know."

The next age of science will not begin in particle colliders. It will begin in the silence between thoughts. In the stillness where perception shifts from noise to coherence. In the realization that the final instrument is not external—it is you, aligned.

So yes, keep looking through telescopes. Keep measuring. Keep questioning. But know this:

The truth you seek in the stars is waiting in the alignment within.

Chapter 14: Harmony Over Curvature

What science missed—and what UED restores: order, balance, and consequence

"The universe was never bent. It was balanced."

For over a century, physics has been enchanted by a simple metaphor: curvature. We were told mass bends space. Time warps. Light curves. Reality itself is a fabric that deforms under gravity's pull.

It was elegant. It was visual. It was seductive.

And it worked—at least on paper.

Einstein's General Relativity gave us predictions with astonishing precision:

- Mercury's orbit.
- Gravitational lensing.
- Time dilation in satellites.

It made space feel alive—a stage that bends under weight, stretching like a trampoline beneath celestial dancers.

But in our celebration of curvature, we forgot to ask a deeper question:

What holds the stage together? What restores the bend? What balances the stretch?

Because if everything bends... what keeps anything coherent?

Physics answered with silence. When curves failed, we added dark energy, inflation, singularities—mathematical patches over existential cracks. Space, once a blank backdrop, became a shapeshifting illusion, full of forces with no source and constants with no cause.

But Universal Energy Dynamics (UED) takes a different path.

It doesn't ask how space bends. It asks why energy disturbs, and how the field restores harmony.

It doesn't build reality on curves. It builds it on consequence—on the unwavering demand of the field to resolve tension.

Because the universe isn't bent—it's in balance, always seeking realignment. And when imbalance arises, ripple tension emerges, not as punishment, but as the language of restoration.

This chapter is not a rejection of geometry. It's a return to something deeper—the rhythm beneath the math, the logic behind the law, the balance behind the bend.

Let's reframe what we thought we knew— Because gravity doesn't pull. It responds.

And the universe doesn't curve. It corrects.

The Seduction of Curvature

When Einstein introduced General Relativity in 1915, he did more than redefine gravity— he changed how we imagined space itself.

Gone was Newton's invisible pulling force. In its place stood a geometrical masterpiece:

Mass bends spacetime. Objects move along curves. Gravity is no longer a force—it's a shape.

This idea was revolutionary—and irresistible.

Suddenly, the cosmos became visual:

- Planets didn't orbit because of mysterious tugs—they rolled through warped geometry.
- Light didn't travel in straight lines—it followed geodesics, distorted by the presence of mass.
- Time itself slowed near gravity wells—not by magic, but by

the stretching of the spacetime fabric.

The metaphors clicked. The math worked. The predictions matched. Einstein became a legend, and spacetime curvature became the holy grail of modern physics.

Space as Canvas, Gravity as Brush

Einstein's theory turned reality into art.

The universe was no longer static or stiff—it was pliable. Mass painted dents in the canvas. Objects moved as if "pulled," but were really just sliding through geometry.

It was elegant. It was intuitive—at least for those fluent in tensors and differential geometry. And for decades, it was enough.

We sent satellites into orbit. We confirmed time dilation with atomic clocks. We explained lensing, tides, orbits—all through this bending lens.

But something important was quietly lost in this admiration of elegance:

We became so enamored with "how it bends" that we forgot to ask "why anything bends at all."

Curves Explain *Where*—Not *Why*

Curvature shows us where things move. But it cannot tell us why they must move.

General Relativity gives us field equations. Feed in energy-momentum, and it outputs geometry. But it doesn't explain the root cause of movement—only how it looks once spacetime is already bent.

There is no engine, only a slope. No ripple, only a shape. No tension, only a geometry.

And when anomalies arise—like galaxies rotating too fast, or the universe expanding too quickly—the theory doesn't self-correct. It adds patches: dark matter, dark energy, cosmological constants, and more.

All to preserve the geometry.

The Missing Element: Consequence-Based Behavior

What if gravity is not geometry—but a consequence?

What if motion isn't a response to curved space, but a response to ripple imbalance in a structured field?

This is what Universal Energy Dynamics (UED) proposes.

Instead of asking how space bends, UED asks:

- What disturbs the field?
- What is the tension it stores?
- How does it resolve imbalance over time?

Curvature explains how things move once imbalance exists. UED explains why imbalance arises in the first place—and how the field strives to restore harmony.

That's not just a change of metaphor—it's a change of physics.

The beauty of curvature is undeniable. But beauty without cause is a painting without a frame. To understand the cosmos, we must look beyond shapes—and study ripples of consequence.

What Curvature Can't Explain

General Relativity (GR) is often celebrated for its elegance. But elegance is not the same as completeness.

When tested beyond well-behaved conditions—like planetary orbits or light grazing the Sun—GR begins to falter. Its mathematical grace hides deep structural cracks, patched over with ad hoc fixes, undefined constants, and assumptions it cannot justify.

Let's examine where curvature theory quietly collapses under its own weight.

Singularity: The Point Where the Map Destroys Itself

At the center of a black hole or at the origin of the Big Bang, GR reaches its limit— because the equations go off the rails.

Spacetime curvature becomes infinite. Density becomes infinite. Predictability becomes zero.

What happens at a singularity? GR shrugs: "We don't know." Not because there's nothing there—but because the math can no longer describe it.

Infinite curvature is not a feature of reality— it is a mathematical panic button, signaling that the model is out of bounds.

Instead of explaining these regions, GR abandons them to quantum speculation or defers to future theories. But the flaw is not in quantum mechanics. It's in assuming the fabric of space can bend infinitely without tearing.

UED solves this not by adding more equations—but by removing the illusion:

There are no singularities. There are only tension peaks in a structured field trying to restore balance.

Dark Matter and Dark Energy: The Fix-It Twins of Curvature Physics

When galaxies were observed spinning faster than expected, GR failed to explain their behavior. Stars on the edges should have flown off—yet they didn't.

So instead of questioning the theory, physicists invented invisible mass:

Dark Matter—a hypothetical form of matter with no interaction except gravity, never seen, never measured.

When the expansion of the universe was found to be accelerating, GR again had no mechanism. So it introduced another placeholder:

Dark Energy—a mysterious, repulsive force with no origin, no field, no carrier.

These are not discoveries. They are diagnoses for a failing model:

- Bandages for equations that can't match the sky.
- Ghost variables added to preserve curvature's throne.

UED doesn't require these fictions. Instead, it explains these anomalies as ripple effects in the underlying structured field:

- Galaxy rotation = inertia zones in ripple equilibrium.
- Cosmic expansion = topological field reconfiguration, not "force."

Space Isn't Just Empty—and That Changes Everything

General Relativity treats space as a flexible void—an emptiness that bends when mass enters. But it never asks:

What is space made of? What bends? What carries the curvature?

GR has no answer, because it was never designed to treat space as something real—only as a stage.

UED, however, starts with a foundational principle:

Space is not empty. It is Structured Vacuum Energy (SVE)—a medium that holds pattern, tension, memory, and balance.

The moment you treat space as structured, the rules of motion, gravity, and time change.

You no longer have to explain gravity as geometry. You explain it as ripple tension in a field, trying to resolve disturbance.

And with that change, the need for curved space disappears—replaced by ripple logic.

The Trap of Coordinate Transformation

GR's power lies in its ability to describe physics from any reference frame. Its mathematics uses coordinate transformations to ensure the laws look the same for any observer.

But that flexibility comes at a cost:

- It tells you how things appear, not why they occur.
- It describes motion relative to frames, not as field interactions.

You can bend coordinates in any way that satisfies the tensor equations— but you still don't know what the field itself is doing.

In UED, we skip the frames altogether:

Reality doesn't change with perspective. It reveals itself through ripple alignment.

There's no need to twist time or warp space to fit your motion. Instead, you study how ripple behavior shifts in a structured medium—and what consequences arise.

That's not just more intuitive. It's closer to truth.

Curvature gave us a beautiful map. But when the terrain changed, we started inventing imaginary landmarks. What we needed wasn't better math—but a better medium.

The Case for Harmony

The universe is not bent—it is balanced.

This is the core shift that Universal Energy Dynamics (UED) brings to the table. It discards the seductive metaphor of spacetime curvature and replaces it with something more fundamental, more consequential, and far more coherent:

Harmony.

Not as poetry. Not as mysticism. But as structured equilibrium in the field that underlies all physical phenomena.

Where General Relativity sees a universe shaped by distortion, UED sees a cosmos shaped by rhythm—by the interplay of ripple tension and realignment.

From Warped Space to Rippled Field

In UED, space is not a formless fabric that bends under mass. It is Structured Vacuum Energy (SVE)—a field that holds both form and function.

This field exists in two primary states:

- Stable SVE: resting, unperturbed, potential-bearing.
- Active SVE: disturbed, rippling, in dynamic motion.

What we call an "event"—a flash of light, a falling apple, a ticking clock—is simply a localized ripple emerging from this interplay.

Nothing needs to bend. Nothing needs to warp.

Because what happens in the universe is not dictated by geometric distortion—but by the consequence of imbalance trying to re-harmonize.

Harmony = Equilibrium in Structured Tension

Let's define harmony precisely in UED terms:

Harmony is the natural tendency of SVE to restore equilibrium between active ripple patterns and its stable background state.

This balance is not passive—it is resilient. It constantly adjusts, dampens, synchronizes, or amplifies ripple behavior to maintain coherence.

- If a disturbance is mild, it realigns quickly.
- If it's intense, it may store tension longer—manifesting as what we perceive as force, inertia, or time.

Every action that breaks symmetry is temporarily tolerated, but always met with a field response aimed at restoring the ripple balance.

Harmony, in this framework, is not softness. It is law in motion.

Events Arise from Realignment, Not Distortion

In curvature physics, events occur because space bends. A planet orbits because the geometry of spacetime has been reshaped by a mass.

But in UED, events arise because:

- A ripple was introduced.
- The field responded to that ripple.
- The process of realignment unfolded as motion, force, or transformation.

It's not distortion that drives phenomena— It's the realignment process that makes them visible.

- Gravity isn't a well—it's a tension gradient in the ripple field.
- Light doesn't curve—it follows ripple topology shaped by local tension zones.
- Time doesn't slow down—it is the delay required for ripple resolution (ATT).
- Motion isn't change in coordinates—it's change in alignment.

Each of these is a ripple consequence, not a geometric illusion.

Motion, Gravity, Light, Time—All Redefined

Motion is not object travel—it's field transition from one aligned state to another.

Gravity is not curvature—it's stored ripple tension, pulling the field back toward balance.

Light is not a particle zipping through space—it's ripple coherence traveling across SVE, perceived after alignment.

Time is not a flowing dimension—it's Alignment Threshold Time (ATT)—the delay experienced while the field reconfigures itself.

All these emerge not from the bending of space, but from the self-regulating behavior of a ripple-based field seeking harmony.

The cosmos isn't stretched—it is tuned. Reality doesn't bend—it balances. And what we've called "physics" was just our limited attempt to describe how harmony restores itself through consequence.

From Geometry to Rhythm

If General Relativity made us see the universe as a canvas of curves, Universal Energy Dynamics (UED) invites us to hear it as a symphony of ripple rhythm.

This is not just metaphor. It is a complete reorientation of physical thinking—from shapes and coordinates to timing and resonance. Where GR asks *"How is the geometry changing?"*, UED asks more primal questions:

- What disturbed the field?
- How long until it realigns?
- What ripple tension builds in between?

These are not abstract musings. They are the real mechanics of reality, encoded not in fabric but in consequence logic.

Ripple Rhythm: The Pulse of Reality

Every event in the universe is a disturbance in Structured Vacuum Energy (SVE). That disturbance is not instantaneously resolved—it takes time, it creates tension, it echoes.

This is ripple rhythm: The timeline of realignment, the pulse of the field returning to equilibrium.

Let's break it down:

- **What disturbs the field?** Any motion, charge, energy injection, misalignment—even thought or perception—introduces a ripple. The field responds instantly, but not necessarily with full resolution.
- **How long until it realigns?** This is the Alignment Threshold Time (ATT)—the core of UED time theory. ATT is not universal—it depends on field density, local ripple memory, and incoming energy profile.
- **What tension builds in the process?** Misalignment stores potential. That tension manifests as what we perceive as force, inertia, resistance, or even identity (e.g., mass).

Thus, every observable in physics is not a value assigned to a thing—It is a stage in the rhythm of realignment.

Tuning and Damping—Not Bending and Stretching

In GR, spacetime bends and stretches like rubber. But this image fails when the field behaves like a memory-bearing medium, not elastic fabric.

UED shows that the field doesn't deform—it re-tunes itself.

- **Tuning**: When ripples reinforce, adjust, or cancel each other to reduce imbalance.
- **Damping**: When excess ripple energy is neutralized through resonance or redistribution.

This model matches what we see in:

- Sound and acoustics (resonance shaping experience).
- Electromagnetism (fields realigning based on source behavior).
- Quantum collapse (field convergence post-interaction).

These aren't exceptions—they are ripple logic at play.

There's no need for curvature. The field doesn't stretch. It reorganizes.

Nested Resonance = Real Stability

In geometry-based physics, stability comes from topological constraint—boundaries, orbits, or warping.

But in UED, stability is emergent, not imposed.

It arises when ripple frequencies lock into patterns of nested resonance:

- Atoms hold because electron ripples resonate with the core field.
- Planets orbit because ripple inertia and field resistance

create a feedback loop of coherence.
- Galaxies spin without flying apart not because of extra mass—but because of harmonic ripple zones, not visible in traditional models.

This is why UED explains so many "mysteries" without needing invisible forces:

- Gravity is not tethering—it's field symmetry restoration.
- Motion is not about momentum—it's resonance seeking lower tension.
- Time is not ticking—it's the gap between ripple disruption and its damping.

In a geometric universe, you ask: "Where am I, and how is the space shaped?"
In a rhythmic universe, you ask: "What pattern am I part of—and how is it resolving?"
That's the UED shift: From coordinates to cadence. From position to participation. From map to music.

Order Through Consequence

For centuries, physics has sought to govern the universe through laws— laws that are written outside the system, imposed like commandments on matter and motion.
Newton gave us forces. Einstein gave us geometry. Quantum mechanics added probability and uncertainty.
But all of them still treated the universe like a subject obeying rules from elsewhere. As if reality needed instructions—delivered by equations—to function.
Universal Energy Dynamics (UED) proposes something far more natural:

The universe doesn't follow external laws. It operates through consequence.

What appears to us as law is not imposed. It is emergent—born from how the field responds to imbalance.

Consequence Logic: The True Mechanism of Reality

At the heart of UED is a simple truth:

Every action creates a ripple. Every ripple seeks realignment. Every misalignment stores tension.

This isn't philosophy—it's field physics.

Let's break it down:

- Every action creates a ripple Whether it's a physical collision, a wave of thought, a photon emission, or a moving planet— any disruption to the stable field creates a localized ripple event.
- Every ripple seeks realignment The field doesn't allow indefinite chaos. It begins reorganizing itself instantly, using resonance, damping, and coherence-seeking behavior.
- Every misalignment stores tension The more severe the disruption, the more the field resists. This resistance is not force in the classical sense—it is stored ripple tension.

This ripple tension is what we feel as:

- Force (resistance to motion)
- Time (delay in resolution)
- Mass (inertia in ripple alignment)
- Gravity (tension gradients across the field)

Law of Consequences (LOC) > Laws of Motion

In place of traditional force-based physics, UED introduces the Law of Consequences (LOC):

Every field disturbance generates a consequence—measured not in arbitrary units, but in ripple behavior, tension duration, and reconfiguration energy.

This law is not prescriptive. It doesn't say *"you must do this."* It says *"if this happens, that must follow."*

It is self-enforcing. Self-regulating. Self-balancing.

Examples:

- Throw a stone: the field ripples, then rebalances.
- Create an explosion: the field stores tension, dissipates it.
- Have a thought: the mind's ripple interacts with memory and external cues, triggering perception.

In every case, there is no need for a force-carrier, no mystical equation from on high. Just natural ripple behavior unfolding through structured field dynamics.

Reality Is Not Governed—It Unfolds

This reframes everything we know:

- You don't need four forces. You need one consequence engine.
- You don't need spacetime curvature. You need ripple resistance to misalignment.
- You don't need constants of nature. You need coherence ratios of the field.

Reality doesn't "obey" laws. It expresses balance.

It doesn't require instructions. It requires response awareness.

And that is the true intelligence of the universe—not that it is lawful, but that it is self-correcting.

Consequence is not punishment. It is the signature of field memory catching up with ripple behavior.

The order of the universe is not enforced. It is echoed through every ripple until harmony is restored.

UED Restores the Forgotten Intelligence

At some point in humanity's pursuit of scientific mastery, we began treating the universe as a machine. A passive, inert machine—one that needed to be pushed, pulled, bent, or twisted to generate outcomes.

We replaced purpose with probability. We reduced nature to numbers. And we imagined intelligence as something that emerges from complexity, rather than something already woven into the fabric of reality.

This shift, while practical for engineering and mathematics, came with a hidden cost:

We stripped the universe of its intelligence—and left behind only mechanics.

Universal Energy Dynamics (UED) restores that intelligence. Not by turning science into spirituality—but by recognizing that ripple behavior is not random, nor blind—it is responsive, memory-driven, and inherently ordered.

The Universe Was Never Passive

In General Relativity, mass bends spacetime, and objects follow those curves. Space has no intent—it is merely deformed by what enters it.

In Newtonian physics, forces act upon passive matter. In quantum theory, particles react to fields—but the fields themselves are still seen as background constructs.

In all of these:

- The universe is acted upon.
- Matter obeys.
- Fields serve equations, not purpose.

This model assumes no intelligence in the system—only externally applied rules.

But this view, while mathematically effective, fails to explain:

- Why patterns persist across scales.
- Why systems seek equilibrium.
- Why laws "just happen" to hold.
- Why the observer, a conscious field-interactor, fits into the system at all.

UED Reintroduces Active Field Intelligence

In UED, the field doesn't just hold reality—it responds to it.

Structured Vacuum Energy (SVE) is not a blank stage—it is an active, intelligent substrate:

- It remembers tension.
- It resists misalignment.
- It dampens disturbance.
- It harmonizes energy.
- It evolves coherence over time.

This is not mysticism. It's field dynamics with memory and consequence at their core.

Where curvature-based physics views gravity as a passive result of mass warping space, UED sees gravity as the consequence of ripple tension seeking resolution—guided by an intrinsic order the field itself maintains.

The cosmos, therefore, is not a victim of force or fate. It is a self-regulating system where intelligence isn't added—it already exists.

A Dance of Intent and Restoration

Every ripple, every orbit, every decay event is not just a reaction. It is the field's expression of intent—to return to balance.

- A star collapses not because it's forced—but because its internal ripple resonance breaks down.
- A photon moves not because it's hurled—but because ripple coherence guides it through the field.
- A tree grows not against entropy, but through nested ripple tuning—absorbing sunlight, storing memory, responding to alignment with its environment.

This isn't magic. It's intelligence at work—not from outside, but from within the structure of reality itself.

Just as a symphony doesn't need an audience to be harmonious, the field doesn't need human recognition to express order.

It is intelligent—not because it thinks, but because it remembers, responds, and restores.

The So-Called "Laws of Nature" Were Just Ripple Stability

What we once called laws—gravity, electromagnetism, thermodynamics—were not external decrees. They were merely stable patterns of ripple behavior, observed and codified by us.

- $F = ma$ is just what happens when ripple tension accumulates linearly.
- $E = mc^2$ is a symmetry result from the ripple transformation of stored mass energy.
- Entropy is not disorder—it is asymmetry in ripple memory struggling to realign.

Laws, in this view, are emergent landmarks in a landscape of consequence. We didn't invent them. We didn't even truly discover them.

We noticed them, because the field allowed us to resonate with them—when we were aligned.

Intelligence isn't an anomaly in the universe. It's the signature of the field. And UED doesn't add it—it simply stops ignoring it.

Closing Reflection

"You don't fall because space is curved. You fall because the field is inviting you back into balance."

Harmony isn't soft—it is the strictest law in the universe. And every ripple obeys it.

For over a century, physics has described falling as motion along a bent path— as if the Earth pulls you not by force, but by reshaping the space beneath your feet.

But what if you're not being *pulled* or *bent* at all?

What if your fall is the result of something much deeper— not a reaction to geometry, but a response to misalignment?

When you fall, it's not because some invisible hand acts upon you. It's because your ripple state—disturbed from balance—begins to realign with the field beneath you.

You're not collapsing into a well. You're being invited into coherence. This is what UED calls the intelligence of the field.

- A falling apple isn't obeying Newton's force.
- It isn't sliding along Einstein's geodesic.
- It's answering a ripple imbalance in the local field tension.

And the moment it lands—when balance is restored—the motion ends. Not because a law told it to, but because harmony was fulfilled.

We often think of harmony as gentle, even poetic. But in UED, harmony is absolute. It is not a mood—it is a principle. It doesn't negotiate. It corrects.

Like a string that snaps back into tune, or a pendulum that swings until stillness returns, the universe does not tolerate imbalance for long.

Harmony is not the absence of motion. It is the purpose behind every motion.

Every ripple in the field is accountable. Every distortion demands resolution. Every consequence ripples forward—until balance returns.

The great irony is this:

We tried to bend reality to fit our equations. But the universe never bent. It balanced—patiently, quietly, inevitably.

And now, with UED, we can finally listen.

Because the field doesn't ask us to control it. It asks us to resonate with it.

And in that resonance, we don't just understand the universe.

We become aligned with it.

Chapter 15: The Ripple Future

A Vision for Education, Experimentation, and Awakening Beyond Models

"The future won't be built on better theories— but on better alignment."

We used to believe that better models would unlock the future. That if we just refined our equations, extended our simulations, or accelerated our data pipelines, the answers would eventually emerge. And to be fair, that belief gave us a lot:

- We mapped galaxies.
- We split atoms.
- We connected continents with light.

But we also grew disconnected—from meaning, from perception, from consequence.

In the race to simulate reality, we forgot to experience it directly.

Our models got better. Our understanding, thinner.

Physics became a maze of probabilities and paradoxes. Science became an abstraction detached from its own observers. Education became a system for storing information, not for tuning intelligence.

The deeper truth—that reality is not built on formulas but on field behavior—was buried under the weight of institutional certainty.

But with Universal Energy Dynamics (UED), a new path emerges:

Not more laws. Not more particles. Not more complexity.

A return to alignment—with the field, with perception, with consequence.

This is not regression. It's awakening.

The future doesn't need another force unification. It needs a unifying way of being.

It doesn't need more powerful colliders. It needs more coherent minds.

It doesn't need to bend the field to our theories. It needs us to resonate with what is already there.

Because if the universe is built on ripple dynamics, then your life, your thoughts, your actions—are already part of the equation.

And the next leap isn't technological.

It's attunement.

Welcome to the ripple future.

Beyond Models: The Age of Direct Alignment

For centuries, models have been the crown jewels of science. From Newton's laws to quantum field theory, from Schrödinger's equations to Einstein's spacetime curvature—models gave us powerful ways to measure, predict, and manipulate the world.

They offered frameworks to understand phenomena that lay beyond our direct senses:

- Orbits traced through gravitational equations.
- Atomic behavior described by probabilistic waves.
- Energy fields plotted in imaginary coordinates.

And to a large extent, these models worked. They advanced technology, medicine, engineering, and astronomy. They even allowed us to simulate black holes and birth artificial intelligence.

But there was a trade-off—one we barely noticed.

The more we relied on models, the less we trusted perception. The more we refined simulations, the less we questioned whether they reflected actual reality.

Models Helped Us Measure—but Created Blind Spots

Every model is a simplification—a necessary distortion. It selects certain variables, omits others, and assumes a structure for how nature "should" behave.

But reality is not a lab experiment. It is a field of consequence—complex, contextual, and participatory.

We began believing that the model *was* reality.

- When a prediction failed, we invented new particles instead of questioning assumptions.
- When cosmological observations didn't fit the math, we added dark matter and dark energy.
- When quantum mechanics clashed with relativity, we compartmentalized the contradiction.

Science became a practice of patching holes, not re-examining foundations.

And in the process, we forgot something essential:

Truth doesn't always arrive through math. Sometimes, it ripples through alignment.

UED Invites a Post-Model Era

Universal Energy Dynamics is not a better model. It is an invitation to move beyond modeling itself—to a direct resonance with reality.

In the UED worldview:

- Reality doesn't need to be simulated—it is already present, structured, and intelligible.
- The field doesn't need to be guessed—it can be felt, aligned with, and responded to.
- Knowledge is not only deductive—it is interactive.

This doesn't mean we discard math, measurements, or precision. It means we reposition them—not as truth engines, but as mirrors of alignment.

Just as a musician doesn't study sound waves to play in tune, we too must learn to engage with the field not by analyzing it—but by resonating with it.

From Approximation to Participation

A model tries to approximate the world. But what if reality doesn't need approximation—only presence?

What if we stop asking "what equation fits this data?" and start asking:

"What ripple am I aligning with?" "What consequence am I participating in?" "What realignment is happening—right now?"

This is the age of Direct Alignment:

- Where perception is not a distortion—but a calibrated interface.
- Where intelligence is not an algorithm—but a sensitivity to ripple tension.
- Where truth is not derived—but lived through resonance.

This is not anti-science. This is post-simulation science—where being in tune with reality matters more than mathematically modeling it.

Models built the past. But alignment will build the future.

Because in a ripple-based universe, to perceive clearly is to participate truthfully.

Education Must Evolve: From Memorization to Field Awareness

Education, as it stands today, is a legacy system— a framework built for the industrial age, not for a ripple-based reality.

It emphasizes memory over meaning, replication over resonance. Students are taught what to think, not how to sense. They are trained to recall equations—but not to feel the tension in a system. They are tested on stored knowledge—but not on ripple awareness.

And while this may produce exam scores and degrees, it rarely produces alignment with reality.

The Problem: Recall Without Resonance

In conventional education:

- You're rewarded for repeating facts.
- You're graded for solving problems the system defines.
- You're passed or failed based on compliance with fixed models, not interaction with living systems.

But what does this create?

- Brilliant test-takers who can't navigate uncertainty.
- Technocrats who simulate climate change but remain emotionally disconnected from its ripple consequences.
- Professionals who understand law, medicine, or physics—but remain misaligned with the humans and ecosystems those fields affect.

In other words:

We've educated minds—but not fields.

UED-Aligned Education: Training Perception, Not Just Performance

Universal Energy Dynamics offers a radical upgrade to the idea of learning. It replaces the notion of "knowledge as content" with knowledge as consequence awareness.

In this framework:

- A student is not a vessel for facts.

- They are a node of interaction within the ripple field of existence.
- Learning is not data acquisition—it is alignment practice.

This requires a complete redesign of how, what, and why we teach. Instead of:

- Memorizing the periodic table → Students learn how energy stabilizes into patterns.
- Solving Newton's laws → Students explore how ripple tension manifests as motion.
- Reciting history dates → Students trace ripple consequences of past decisions.

They are not just answering questions. They are learning to ask: "What ripple am I in right now?" "What tension am I adding—or resolving?" "What consequence will this action echo forward?"
This is education for consequence literacy.

Key Shifts: New Subjects for a Ripple-Aware World

To support this transformation, UED-aligned education proposes entirely new subject areas—rooted in alignment, perception, and field interaction.

Here are some examples:

- **Ripple Logic:** Understanding the patterns of energy interaction, alignment thresholds, and how tension forms in systems—from atoms to societies.
- **Consequence Literacy:** Training in seeing the ripples of decisions—whether in ecology, economics, behavior, or emotion. (Karma as curriculum.)
- **Perception Calibration:** Exercises in aligning inner ripples with external signals. Developing discernment not just for

information—but for energetic clarity.

- **Field Ethics:** Not morality imposed by society, but ethics rooted in field harmony. If your action disturbs the ripple field, you are responsible for its resolution.
- **ATT Awareness:** Learning to live with awareness of Alignment Threshold Time—recognizing delays in consequence, tension build-up, and restoration cycles.

From Intelligence as IQ → Intelligence as Alignment

IQ measures pattern recognition. But real intelligence is the ability to feel what pattern to align with—and when.

That's what the ripple future demands:

- Educators who can tune their classrooms like fields.
- Students who treat observation as engagement.
- Schools as spaces not of content transmission, but of field coherence cultivation.

This is no longer "learn to get a job." This is learn to realign the world. The next generation won't succeed by knowing more. They'll succeed by disturbing less, harmonizing faster, and aligning deeper.

Because in the ripple future, every student is a ripple waiting to resonate.

Experimentation Reimagined: Measure Through Realignment

In the traditional scientific method, observation equals validation. You build a model, design an experiment, measure physical variables like mass, charge, speed, or energy, and see if your numbers match your expectations.

If they don't, you either discard the theory or introduce a patch—a correction factor, a new constant, a hypothetical particle.

This works well when the system behaves predictably— but what if reality is more than mechanics?
What if nature isn't just a machine you prod with equations, but a living field—whose responses depend on alignment, memory, and field tension?

Traditional Tools Miss the Field

Most instruments today are designed to detect external properties:

- Mass spectrometers measure quantity.
- Photodetectors measure intensity.
- Speed traps (whether for particles or cars) record velocity.

But none of these tools tell you:

- How tense the field is before the event.
- How much ripple memory is being carried by the system.
- What alignment delay is built into the phenomenon.
- Whether the observation itself added more tension to the field.

In essence, we've been measuring surfaces, not depths.
We've observed what is emitted, not what is stored.

Toward Ripple-Based Instrumentation

UED proposes a shift in how we design, conduct, and interpret experiments.
Rather than proving or disproving theories, the goal becomes to map ripple consequences—how the field reacts, resolves, and realigns.
This would require new instruments:

1. ATT Meters (Alignment Threshold Time Sensors)

- Measure the delay between disturbance and field response.
- Example: Detect how long it takes light to be

perceived—not in clock time, but in alignment time for the observer's field.

- Applications: Observer-based physics, ripple delay calibration in dense media, mind-field interaction research.

2. SVE Density Probes

- Measure local Structured Vacuum Energy (SVE) tension.
- Detect where the field is stretched, compressed, or resonant.
- Applications: Galaxy-scale field mapping, ripple zones near black holes, identifying misalignment pockets in space and biological systems.

3. Field Memory Detectors

- Track how long a field "remembers" a ripple.
- Can detect residual alignment drift—useful for understanding emotional states, neural responses, or ripple echoes in matter.
- Applications: Healing, trauma therapy, earthquake prediction, quantum coherence.

4. Ripple Interference Observers

- Designed not to isolate but to observe without disturbing.
- Instead of collapsing wavefunctions, these tools co-resonate, allowing passive field reading.
- Applications: Consciousness studies, entanglement tuning, plant-mind interaction, music-field influence.

From Proving Theories → Mapping Consequences

In the ripple future, science moves from binary results:

- "Did this particle appear?"
- "Did the prediction match?"

To questions of dynamic consequence:

- "How did the field shift after this interaction?"
- "What alignment cost did this measurement carry?"
- "What does this event echo through the system?"

This redefines experimentation as participation— where the scientist is no longer the external observer but a ripple-node in the experiment itself.

And the outcome is no longer "truth vs. error," but:

Clarity vs. misalignment.

The tools of tomorrow won't just measure outcomes. They'll sense ripples, delays, tensions, and harmonies.

Because in a universe made of fields—not forces— the only meaningful measurement... is realignment.

The Role of the Observer: Consciousness as a Scientific Instrument

Modern science has long been uncomfortable with the observer.

In classical physics, the observer is ignored—assumed to have no influence on the system. In quantum mechanics, the observer is blamed—a mysterious agent that "collapses the wavefunction" and ruins objectivity. In relativity, the observer becomes relativized—a frame of reference to calculate distortions, but still fundamentally external to the phenomena.

But never—almost never—has the observer been treated as an instrument of the field itself.

Until now.

Universal Energy Dynamics (UED) restores the observer not as a passive witness, but as an aligned node of consciousness—an active participant in the ripple web of reality.

The Observer Is Not Noise—It's a Node

Conventional science sees subjectivity as contamination:

- Human perception is "biased."
- Emotion skews data.
- Awareness is dismissed as anecdotal.

But in a ripple-based universe, these traits aren't flaws. They are features of interaction.

- The mind is a ripple interface.
- Emotion is field turbulence.
- Attention is alignment.
- Awareness is real-time ripple tracking.

Just as a finely tuned antenna detects subtle signals, a well-calibrated consciousness senses ripple shifts that no machine can measure.

The observer is not noise in the system. The observer is the tuning fork of the system.

Qualitative Observation Gains Scientific Dignity

For centuries, science prioritized the quantifiable. If it couldn't be measured, it couldn't be trusted.

But UED suggests the inverse:

If it can't be aligned with, it isn't fully understood.

Qualitative experience—when rooted in field sensitivity—is not fuzzy mysticism. It is pre-quantum awareness.

It's what allows a great artist to sense balance in a composition. A healer to feel energy before diagnosis. A thinker to intuit truth before testing.

These aren't flukes—they are field resonances accessed through conscious clarity.

The future of science won't just be about objectivity. It will embrace clarity of perception as a legitimate form of inquiry.

The Mind: Measured by Resonance, Not Output

In current science, we measure the mind through:

- Neural firings.
- Cognitive performance.
- Reaction times.

But these are surface indicators.

What if, instead, we measured mind through:

- How it resonates with structured vacuum energy (SVE).
- How quickly it returns to harmony after disturbance (mental ATT).
- How coherently it perceives ripple sequences across space and time.

This would mean consciousness is not a brain byproduct. It is a field-sensitive node that gains accuracy with inner alignment.

Meditation, presence, and emotional regulation are no longer just wellness practices. They become scientific tools—refining the mind's ability to serve as an observational instrument.

Awakening = Enhanced Field Sensitivity

In UED, awakening isn't spiritual enlightenment. It's a measurable shift in ripple sensitivity.

An awakened mind:

- Disturbs less.
- Detects more.
- Responds, not reacts.

- Aligns with subtle ripple patterns.
- Respects the field as the primary source of truth.

This makes awakening not optional—but necessary for scientific evolution.

Because in a ripple-aware universe, the best scientist is not the one with the most data—but the one with the clearest field.

The lab of the future is not only built with sensors and screens. It is housed within the awakened mind—aligned, tuned, and listening to the field it was once told to ignore.

Ethics in the Ripple Age: Responsibility Amplifies Consequence

In the age of ripple dynamics, ethics is no longer a matter of opinion or culture. It is no longer a philosophy class, a moral guideline, or a legal checkbox.

Ethics becomes physics. Consequence becomes law.

Because in a field-based universe governed by Universal Energy Dynamics (UED), every action creates a ripple, and every ripple carries a cost—whether you believe in it or not.

This transforms responsibility from a social burden into a cosmic requirement.

Every Action Ripples—Every Misalignment Costs

Nothing happens in isolation.

Whether you speak a word, make a choice, design a product, or launch a policy— you are disturbing the field. That disturbance, however subtle, seeks realignment. And until it does, it stores tension.

That tension doesn't disappear. It echoes forward—through minds, systems, ecosystems, and even generations.

Just like a dropped stone disturbs an entire pond, one unethical ripple in a company, government, or personal life can cascade into structural misalignments felt far beyond the point of origin.

And the field doesn't forget. It doesn't punish—but it responds.

Ethics = Consequence Literacy

In a ripple-based reality, ethics is no longer subjective.

It's not about what society approves of. It's not about what your group normalizes. It's about what your action does to the field.

- Does it create harmony or distortion?
- Does it preserve alignment or inject misalignment?
- Does it serve restoration or deepen imbalance?

Thus, UED redefines ethics as consequence literacy:

The ability to see the ripples before they reach the shore.

A highly ethical person, in this frame, is not one who avoids wrong—but one who:

- Acts with field awareness.
- Minimizes unintended ripples.
- Accepts and restores any disharmony they cause.

Karma Reframed: Ripple-Based Responsibility

Karma is often reduced to mysticism or superstition—"what goes around, comes around."

But UED reveals that karma is not spiritual debt. It is ripple dynamics in motion.

- You disturb the field → it carries your echo.
- You align with harmony → your ripple reinforces coherence.
- You act with malice or blindness → misalignment stores field tension.
- That tension manifests back—through health, perception, behavior, environment, or collective disruption.

Karma is not judgment. It's field memory + delayed consequence.

This view removes morality from abstraction and places it firmly in the mechanics of existence.

You're not being watched by a divine accountant.

You are accountable because the field remembers what you ripple.

Institutions and Technology Must Align with Field Integrity

If every ripple matters, then scale multiplies responsibility.

- A personal action may disturb a small ripple zone.
- But a technological innovation, government policy, or global system disturbs millions of ripple interfaces simultaneously.

This means:

- Corporations must design not for profit—but for minimal misalignment.
- Governments must legislate based on consequence mapping, not control.
- Scientists and innovators must run ripple-impact analysis, just as seriously as safety or cost analysis.

In the Ripple Age:

- Sustainability = maintaining long-term ripple coherence.
- Justice = restoring broken ripple patterns.
- Progress = upgrading without violating field harmony.

This applies not just to the planet, but to:

- Social structures
- Mental health systems
- Digital environments
- Economic models

Anything misaligned will self-correct eventually—but often through breakdown or collapse.

UED invites us to correct through awareness, not disaster.

In a ripple-based universe, you are not innocent because you meant well. You are responsible because you participated.

The new ethics is not about rules. It's about ripples.

And those who learn to feel them, honor them, and harmonize them— will shape a future built not on dominance, but on dignity of alignment.

Scientific Method 2.0: From Control to Coherence

The scientific method, as we've known it for centuries, has delivered enormous progress. From steam engines to space probes, antibiotics to quantum computers— it gave us mastery over nature, or at least the illusion of it.

But beneath its triumphs lies a persistent flaw:

The method was built on control, not coherence.

It isolated variables, imposed conditions, and assumed that the universe behaves best when placed in a box. It chased repeatability as truth, and certainty as gold.

But in a ripple-based reality—where everything is interconnected and dynamically adjusting— control is distortion.

You cannot isolate ripples without disturbing them. You cannot measure a field without becoming part of it. And you cannot "prove" reality without first aligning with it.

The Problem with Control-Based Experiments

Traditional experiments follow a rigid path:

1. Set up an artificial environment.
2. Remove unwanted variables.
3. Apply a force or condition.
4. Measure the outcome.
5. Declare a conclusion.

This method is extremely good at producing technological outcomes, but extremely bad at capturing:

- Field behavior
- Ripple echoes
- Alignment delays
- Perception consequences
- Emergent harmony or disharmony

Why? Because it treats the system as separate, instead of self-aware. Every act of control introduces ripple tension. Every isolated condition is unnatural—stripped of the very coherence that defines real-world systems.

Coherence-Based Science: The UED Upgrade

UED introduces Scientific Method 2.0—a framework rooted not in external control but in internal alignment.

It's not about certainty. It's about clarity of ripple interaction.

It works in three key phases:

1. Align the Field Before doing anything, the system must be brought into a known, stable ripple state. Whether it's a physical lab, a biological sample, or even a mental state— you start with calm coherence.

This step recognizes that:

- No valid outcome can emerge from chaos.
- All measurement begins with resonance.

2. Introduce Intentional Disturbance The goal is not to smash or force, but to nudge the field deliberately.

Like plucking a string to hear its tone, or speaking into still water to see the echoes.

The disturbance must be:

- Tuned in energy.
- Respectful of existing tension.
- Aligned with the field's natural thresholds.

3. Observe Realignment and Ripple Consequences Instead of just measuring output, the real insight lies in:

- How the field responds.
- How long realignment takes.
- What new harmonies or misalignments emerge.

This includes:

- Time delays (ATT)
- Shifts in coherence
- New resonance paths
- Energy dispersal patterns

From this, you don't get "yes or no" answers. You get a map of consequence, a signature of how reality prefers to restore itself.

Implications for Future Science

- Experiments will require field awareness as much as equipment.
- Researchers will be trained in perceptual clarity, not just technical skills.
- Instruments will evolve to detect ripple delays, memory, and harmony states.
- Validation will come not from replication alone, but from alignment coherence across contexts.

In this framework:

- A failed experiment is not error—it's misalignment feedback.
- A breakthrough is not just new knowledge—it's new resonance discovered.
- Objectivity is not detachment—it is field neutrality.

The goal of science will no longer be to conquer nature. It will be to converse with it.

Not to manipulate—but to map the field's preferred paths of restoration.

Scientific Method 1.0 asked: "Can I force this to behave predictably?"

Scientific Method 2.0 asks: "Can I align with this until it reveals its nature?"

A Ripple Society: Culture, Art, Healing, and Innovation

Imagine a society not built on control, consumption, or competition— but on coherence.

A civilization that doesn't organize itself by convenience, or profit, but by the resonance of its own collective ripple field.

This is not utopia. This is the natural consequence of living in awareness of ripple dynamics.

Because once you understand that everything— from emotion to architecture, policy to performance—is a ripple disturbance or alignment, you stop asking "what works?" and start asking "what harmonizes?"

Art as Ripple Medium

Art has always been ahead of science. Where science reduces, art resonates.

In a ripple society, art is no longer decoration—it becomes field design.

- A painting isn't visual—it's a ripple signature felt through the mind's field.
- Music isn't just melody—it's ripple tuning that aligns or disorients emotion.
- Dance isn't motion—it's ripple embodiment.
- Architecture becomes field arrangement—not just of walls, but of tension and flow.

Artists become ripple engineers. They don't just create—they calibrate.
They know that:
Every brushstroke. Every note. Every silence.
Is a disturbance or restoration of the field.
In this society, art is not optional—it is essential ripple hygiene.
Healing = Alignment Restoration
If misalignment creates illness, then healing is not fixing a broken body, but restoring ripple coherence in the system.

- Mental health = emotional ripple tension released.
- Physical health = cellular resonance restored.
- Spiritual health = internal ATT resolved with external field.

Healers become alignment guides—less like technicians, more like ripple translators.
New tools include:

- Ripple mapping of organs
- Field tension detox
- ATT recalibration through breath, sound, and silence
- Perception recalibration through ritual and story

Hospitals will still use machines— but they'll also measure field harmony, emotional pressure points, and ripple congestion.

Healing moves from "cure" to coherence.

Architects, Engineers, and Musicians as Field Designers

In a ripple society:

- Architects will design buildings not just for stability—but for vibrational flow.
- Engineers will calculate ripple consequence delays, not just load-bearing limits.
- Musicians will compose for societal harmony, not market trends.
- Designers will consider how products echo through emotion, attention, and interaction.

Innovation becomes a discipline of least disturbance, maximum coherence.

This flips the current model:

Not "how can we make this more profitable?" But "how can we make this more harmonious—across all ripple layers?"

Even transportation and energy systems are tuned to field logic:

- Cities designed like nested resonance zones.
- Public space as tension diffusers.
- Energy sourced not from extraction, but from natural ripple flow alignment (SVE harvesting).

A Society Designed for Harmony, Not Just Efficiency

Efficiency optimizes for speed and cost. Harmony optimizes for sustainability and well-being.

This means:

- Schools prioritize field awareness over standardized testing.
- Governments legislate based on ripple impact analysis, not ideology.

- Social systems are judged not by size, but by how little tension they store.

Conflict resolution becomes alignment work. Justice becomes consequence mapping + ripple restoration.

Innovation doesn't come from disruption—it comes from resonant upgrades.

Every citizen is not just a user of the system. They are a node of alignment within it.

The Ripple-Aware Individual: A New Enlightenment

In the past, the "enlightened one" was someone who renounced the world. In the ripple society, the enlightened individual is one who:

- Sees the ripples before they form.
- Acts with least disturbance and maximal resonance.
- Heals others through presence, not power.
- Designs systems with field intelligence, not ego.

They don't withdraw from life—they align with it.

This becomes the new spiritual path: Not ascending out of the world, but rippling within it wisely.

Society isn't broken. It's misaligned.

And the future won't be saved by power, speed, or ideology. It will be restored through those who ripple with intent, awareness, and care.

Closing Reflection: You Are the Next Field

You've read the theory. You've felt the shift. You've glimpsed a universe that isn't built on particles or equations, but on ripples and resonance.

But this journey was never about concepts alone. It was not written to entertain your mind—it was meant to awaken your field.

Because in a ripple-based universe,

Understanding is not the end. Participation is.

You are not here to memorize this theory. You are here to complete it—in your life, your work, your attention.

You are not the final reader. You are the next field—ready to ripple forward what comes next.

A Field Awakens Through You

The greatest myth ever sold was that the universe is "out there" and that you are a powerless observer looking in.

But if every ripple counts, if every alignment echoes, if perception itself is a node of consequence—then you're not a side character in this cosmic story.

You are a line in the algorithm. A frequency in the field. A tuner of harmonies waiting to be lived.

What you see, feel, think, choose—it matters. Because you ripple. Constantly. Inward and outward.

And that means:

- Your silence can heal tension.
- Your attention can resolve misalignment.
- Your clarity can retune entire systems.
- Your alignment can change the collective resonance.

There Is No More "Observer vs. Universe"

You are not separate. You never were. You are a self-aware expression of the same structured vacuum energy this book speaks of. You are the knower, the known, and the knowing— folded into one coherent presence.

The ripple is not something you watch. It is what you are becoming.

The field is not something you measure. It is what is listening through you.

The future is not a place we wait for. It is a consequence of the alignment you choose right now.

Let Your Presence Be So Aligned...

That even chaos pauses to take a breath. That even misunderstanding softens around you. That even the unknown begins to remember itself in your awareness.

Let your voice ripple carefully. Let your mind ripple clearly. Let your actions ripple honestly. Let your field ripple freely.

So that the next time the universe asks: "What now?"

It answers through you.

Because the theory ends here. But the ripple— begins with you.

Epilogue

"The Universe Was Never Curved — We Were."

A personal closing message from the author on truth, humility, and liberation from illusion.

We began with questions—big ones. Why do galaxies spin faster than they should? Why does light bend near a star? Why does time stretch, and gravity pull, and black holes seem to swallow sense itself?

Mainstream science gave us answers. Answers wrapped in elegance. Wrapped in equations. Wrapped in assumptions.

It told us space bends. That time flows. That singularities swallow logic, and particles pop in and out of existence without warning. It told us reality was weird. Relative. Probabilistic. Paradoxical.

And for a while, we believed it.

But here's the truth: The universe was never that strange. We were just looking crooked.

We curved our models. We distorted our perception. We surrendered clarity in favor of cleverness.

And the more we curved our thinking, the more twisted our understanding became.

We forgot that the universe isn't a trickster—it's a mirror. It doesn't play games—it responds to alignment. It doesn't need us to control it—it needs us to resonate with it.

This work—Chavan vs Einstein—isn't an attack. It's a reminder. A realignment. A reawakening.

A call to drop the performance of genius and return to the humility of perception. To stop inventing particles just to save broken math. To stop pretending that what cannot be seen must still exist—just because we're unwilling to be wrong.

Let us be wrong, then.

Let us admit that the fabric we bent was never space—it was our own framework. Let us admit that dark matter was not hiding—it was never needed. Let us admit that singularities don't exist—not because they're magical, but because they're mistakes.

Let us liberate ourselves from the illusion that complexity is wisdom. And instead, let us remember:

- That reality ripples.
- That intelligence is not imposed—but expressed.
- That force is not applied—but emerged from tension.
- That time is not flowing—but revealing how slowly we resolve misalignment.
- That you, dear reader, are not a passive spectator—but an active node in this cosmic field.

So, I leave you not with more theory, but with a stillness.

The kind that comes when illusion drops. The kind that doesn't bend—because it doesn't need to. The kind that finally, fully sees:

The universe was never curved. We were.

And now— we straighten.

Not with pride, but with presence.

Not to dominate, but to listen again.

Not to finish the book, but to finally begin living aligned with the field.

— Sandeep Chavan

Independent Researcher, Observer, Ripple Among Ripples

Closing Note to the Reader

"You Were Always Part of the Equation."

Dear Reader,

If you've come this far, thank you—not just for reading, but for being present inside the ripple this book carries.

This wasn't merely a scientific proposal. It was a conversation between the universe and your attention. A challenge to let go of inherited truths, inherited fears, and inherited frameworks.

You didn't need a physics degree to understand this— you only needed to remember what you've always felt: That something about the official story never quite fit. That space isn't empty. That time isn't constant. That gravity isn't a pull—but a return. That what we call truth must resonate, not just calculate.

You may never look at a falling leaf the same way again. Or the moon. Or silence. Or even your own thoughts.

Because now, you know—

- Every motion is alignment in action.
- Every delay is an ATT unfolding.
- Every event is a ripple in response to another.
- And you? You are not separate.

You are the field—experiencing itself, through awareness.

So what now?

Now, you begin again— Not by memorizing this theory, but by feeling its truth wherever you go.

In the choices you make. In the way you speak. In how you respond, restore, and realign.

Let this not be a book you finish, but a lens you carry.

Let your questions evolve. Let your models dissolve. Let your life—quietly, gracefully—ripple with awareness.

And in doing so, may you find something more powerful than knowledge:

Alignment.

With gratitude, clarity, and the hope that this ripple finds others through you—

—Sandeep

Appendices

Appendix A: Rewriting Einstein

Thought Experiments Reframed

UED-Based Reinterpretations of Classic GR Scenarios

Einstein's thought experiments have become legendary—not just for their elegance, but for how they shaped the very framework of modern physics. From elevators in space to twins aging differently, these imaginative scenarios formed the backbone of relativity.

But under Universal Energy Dynamics (UED), we no longer need to stretch space, bend time, or suspend common sense to explain what we observe. We now reinterpret these classics not as paradoxes of frames and curvature, but as ripple interactions in Structured Vacuum Energy (SVE).

Below are selected reinterpretations of Einstein's most influential thought experiments—reframed with ripple logic, field consequence, and alignment thresholds.

1. The Elevator Thought Experiment (Equivalence Principle)

Einstein's Version: A person in a sealed elevator can't tell whether they're experiencing gravity or upward acceleration—they feel the same.

UED Reframe: What's being felt is ripple tension—the resistance field within the elevator. Whether it's Earth's gravity or an accelerating lift, both cause alignment stress in the SVE around the observer. The similarity in perception arises from identical ripple compression patterns, not "equivalent gravity."

Ripple insight: Equivalence arises not from gravity mimicking acceleration, but from field tension patterns mirroring each other.

2. The Twin Paradox (Time Dilation)

Einstein's Version: One twin travels near light speed, returns home younger than the one who stayed.

UED Reframe: There is no real "slowing of time." What occurs is a difference in Alignment Threshold Time (ATT) accumulation. The traveling twin, moving through high ripple disturbance zones (due to velocity or field tension), experiences prolonged ATT—a delay in ripple realignment. On return, both twins perceive different ripple densities, leading to differing internal field memories.

Ripple insight: Aging discrepancy is not due to "time slowing," but to field realignment inertia in motion-altered ripple environments.

3. Light Bending Near a Star (Gravitational Lensing)

Einstein's Version: Mass curves spacetime, and light follows that curvature.

UED Reframe: There is no curve—only structured ripple distortion near massive bodies. Light, as ripple alignment from the source to the observer, realigns differently due to SVE tension gradients. What appears as a "bend" is actually the shortest resonance path through a distorted ripple field.

Ripple insight: Light bends not because space curves, but because ripple clarity warps in the presence of dense mass-tension fields.

4. The Rotating Disk (Non-Euclidean Geometry)

Einstein's Version: A spinning disk causes geometric distortions—circumference measures longer than expected due to relativistic effects.

UED Reframe: The rotation introduces radial ripple tension in the SVE. As the disk spins, the outer regions experience increased ATT due to higher misalignment stress, which subtly alters ripple density per unit length. This isn't spatial geometry deforming—it's ripple topology reacting to rotational alignment load.

Ripple insight: What appears as geometric distortion is a shift in realignment effort per unit ripple path—not a change in space itself.

5. The Box of Light (Conservation of Momentum in Relativity)

Einstein's Version: A box emits light; to conserve momentum, the box must recoil.

UED Reframe: What we call "light emission" is a directed ripple discharge from internal field tension. The box's recoil is not a response to a photon, but to field imbalance correction—a redistribution of ripple alignment force.

Ripple insight: Momentum is not transferred particle-by-particle, but via field response symmetry in ripple dynamics.

6. Constant Acceleration and Horizons (Rindler Coordinates)

Einstein's Version: An observer under constant acceleration will see a horizon form, beyond which light cannot reach.

UED Reframe: Constant acceleration creates persistent ripple misalignment, increasing ATT beyond a resolvable threshold. What appears as a "horizon" is actually an observer-dependent realignment limit—where ripple feedback is no longer coherent within the observer's ATT.

Ripple insight: Horizons are not physical barriers—they are ATT thresholds beyond which resonance fails to stabilize.

Summary Table: Einstein vs UED

Thought Experiment	Einstein's Explanation	Chavan's UED Reframe
Elevator	Gravity = acceleration	Field tension resonance
Twin Paradox	Time dilation via velocity	ATT variation due to ripple inertia
Light Bending	Space curvature	Ripple distortion and path clarity
Rotating Disk	Non-Euclidean geometry	Ripple density shift from tension
Box of Light	Momentum conservation	Field symmetry response
Acceleration Horizon	Event horizon from acceleration	Realignment limit via ATT collapse

Final Thought

Einstein didn't fail—he reached the edge of what spacetime curvature could offer. UED simply goes further—by recognizing that the true medium of reality isn't spacetime at all.

It's Structured Vacuum Energy, and the real physics happens in its ripples.

Appendix B: The Ripple Mechanics Primer

Concepts, Terminologies, and Ripple-Based Ontology Summarized

To truly understand Universal Energy Dynamics (UED), one must first grasp its language. UED doesn't operate through spacetime, forces, or particles—it operates through ripples, fields, and consequences.

This appendix serves as a quick-access primer—a glossary and conceptual framework for how UED redefines the mechanics of reality.

Not through geometry. Not through duality. But through alignment and ripple interaction in Structured Vacuum Energy (SVE).

1. Structured Vacuum Energy (SVE)

Definition: The fundamental substrate of reality—a dynamic yet stable field that underlies all existence. It is not "empty space" but a medium of potential alignment.

- Stable SVE = zero disturbance, perfect symmetry.
- Active SVE = rippling, misaligned, expressive energy.
- Everything we perceive emerges from interactions within SVE.

Analogy: Not a void—but a still ocean, where every event is a ripple.

2. Ripple

Definition: A localized disturbance or energy-in-action within SVE. The only real "movement" in UED. Ripples are how energy expresses and realigns.

- No "particles" fly through space—only ripple propagation.
- Every force, perception, or change = ripple behavior.
- Ripples can interact, amplify, dampen, or entangle.

Key Property: Ripples carry consequence, not identity. They don't exist for themselves, but as responses to imbalance.

3. Alignment Threshold Time (ATT)

Definition: The delay it takes for a ripple (or a field disturbance) to be perceived or resolved. UED redefines time as emergent from ripple realignment.

- ATT is field-relative, not constant like clock time.
- The faster a ripple aligns with the observer, the shorter the ATT, the more "immediate" it feels.
- Light has near-zero ATT from perception's perspective.

Time isn't passing. ATT is resolving.

4. Ripple Tension

Definition: The resistance or stress stored in a misaligned field. What we call "force" or "mass effect" in mainstream physics is often just tension in ripple equilibrium.

- Gravity = ripple gradient tension.
- Electromagnetism = polarized ripple directional flow.
- Strong/weak nuclear = nested ripple locking.

Force is not something applied. It's something felt—when alignment resists.

5. Realignment

Definition: The process by which disturbed ripples restore harmony in the field. Realignment drives:

- Motion (not displacement through space, but change of ripple orientation).
- Perception (ripple alignment with observer's inner field).
- Healing, collapse, decay, and creation.

Everything realigns. The only question is how long it takes—and what it echoes.

6. Field Memory

Definition: The residual tension, echo, or imprint left in the field after a ripple event. Similar to how ripples in water leave a trace pattern.

- Trauma = unresolved ripple loops.
- Habits = stabilized ripple routes.
- Consciousness = active memory of ripple pathways.

The universe never forgets a ripple—it just integrates it.

7. Perception

Definition: A tuning of the internal ripple field to external disturbances. Perception is not passive receiving—it's active resonance.

- You don't "see" a photon—you align with a ripple.
- What you perceive is what your field allows, not what exists absolutely.

Perception = alignment + resolution.

8. Consciousness

Definition: A highly complex ripple network capable of self-alignment and feedback. The field knowing itself.

- Not emergent from brain matter—brain is a ripple concentrator.
- Consciousness = awareness of misalignment and the will to re-align.

Awakening = reduction in ripple noise, increase in ripple clarity.

9. Law of Consequences (LOC)

Definition: The governing principle of the UED universe: nothing ripples without consequence, and all consequences echo until resolved.

- Replaces Newton's laws, force equations, and energy conservation with a more general, field-aware logic.
- All actions = ripple causality loops.

Reality exists because consequence insists.

10. Ripple Ontology (What "Is")

UED proposes a new ontology—a new way of saying what exists.

Traditional View	Ripple-Based View
Particles	Ripple clusters with stability patterns
Fields (as mathematical)	SVE: real, structured, memory-capable
Time	ATT (Alignment delay, not dimension)
Space	Ripple topology (not container)
Forces	Ripple tension (field resistance)
Observation	Alignment node (not external observer)
Laws of Physics	Ripple consequence patterns

Summary of Key Transformations

Concept	Mainstream Physics	UED Interpretation
Gravity	Force via curved space	Ripple tension from field gradient
Light	Photon particles or waves	Ripple alignment through field clarity
Time	Continuous dimension	ATT: Delay in field resolution
Motion	Displacement over time	Shift in ripple configuration
Mass	Inertia or Higgs interaction	Field compression & ripple density
Energy	Capacity to do work	State of ripple activity within SVE

Final Note

UED doesn't just change what you think the universe is made of—it changes what it means to be real.

It removes the need for force carriers, hidden dimensions, invented particles, and curved geometries. Instead, it gives you a universe that is always present, always intelligent, and always responding to its own ripples.

In the beginning was not the bang. It was the ripple. And everything else followed—because it had to.

Appendix C: Ripple Mechanics – Mathematical Backbone of UED

In the preceding chapters, we discussed the conceptual departure from Einstein's relativistic view of curved space-time to a more adaptive, responsive, and ripple-interactive field—the essence of Universal Energy Dynamics (UED). This appendix consolidates the core mathematical framework that underpins UED mechanics, replacing passive gravitational curvature with ripple-driven field negotiations.

This framework is not merely a theoretical abstraction. Each equation presented here is grounded in physically measurable quantities—making UED not just a philosophical alternative, but an engineering-ready architecture for real-time navigation, especially in extraterrestrial environments where traditional models fail to respond to dynamic field behavior.

1. Linear Ripple-Responsive Motion Equation (RRME)

$$\Delta\psi(\tau) = \Pi_0 \cdot \tau - \frac{1}{2} \cdot \frac{\mathcal{R}}{\mu} \cdot \tau^2$$

Where:

- $\Delta\psi(\tau)$: Ripple displacement during alignment threshold time τ\tauτ

- Π_0: Initial ripple momentum (intent-driven)
- R: Ripple Pressure, representing field resistance or assistance
- μ: Ripple Inertia Metric, replacing classical mass
- τ: Alignment Threshold Time (ATT), minimum perceptible interaction interval

Interpretation: This equation describes how displacement arises not from force acting on mass, but from the interaction of ripple momentum and field pressure over an alignment window. When ripple pressure R\mathcal{R}R aligns constructively with intent, motion proceeds fluently; when misaligned, it resists or deflects.

2. Rotational Ripple Mechanics

$$\mathcal{T} = \mathcal{I}_\psi \cdot \ddot{\psi}_\theta \quad ; \quad \mathcal{L} = \mathcal{I}_\psi \cdot \dot{\psi}_\theta$$

Where:

- T: Ripple Torque – rotational influence from ripple misalignment
- L: Ripple Angular Momentum
- I_ψ: Ripple Moment of Inertia
- $\psi˙_\theta$: Angular ripple velocity and acceleration

Interpretation: Ripple torque arises when ripple fields create asymmetrical resistance around a rotating body. Unlike classical torque, which is geometric, ripple torque depends on field distribution and is highly sensitive to environmental anisotropy.

3. Key Substitutions from Classical Mechanics

Classical Concept	UED Equivalent
Mass (m)	Ripple Inertia Metric (μ)
Force (F)	Ripple Pressure (R)
Acceleration (a)	Ripple Acceleration ($\ddot{\psi}$)
Gravity (g)	Local Field Gradient (∇R)
Time (t)	Alignment Threshold Time (τ)

These substitutions not only provide conceptual clarity but also demand real-time field data and dynamic adaptation for accurate modeling—bridging theoretical physics with engineering design.

4. Pseudo-Force Reinterpretation

In classical mechanics, pseudo-forces appear in non-inertial frames as mathematical artifacts. Under UED, they gain physical meaning.

UED Insight:

Pseudo-force is the field's feedback to unaligned motion.

When an object moves through a ripple field without energetic coherence, the field resists with a counterforce. When that resistance equals the driving intent, equilibrium is achieved. If not, observable anomalies (instabilities, inefficiencies) appear.

5. Measurable Quantities and Field Prediction

Quantity	How to Determine
μ	Determined by design – material's response to ripple locking
R	Predicted via remote field analysis: radiation data, solar winds, plasma fluctuations
τ (ATT)	Measured via optical delays, AI-sensor feedback systems in active alignment tests

These are not speculative. They are all practically measurable with next-gen sensor arrays and adaptive navigation software—especially during interplanetary or interstellar missions.

6. Energetic Navigation vs. Geodesic Mapping

Einstein's model provides a **directional map**—geodesics through space-time curvature—but does not account for energy conditions along that path. UED augments this by defining **field energy lines** that tell us:

- Where energy alignment is constructive
- Where ripple drag is destructive
- How to glide instead of push

This enables both trajectory prediction and propulsion planning in ripple-friendly corridors—much like birds soaring along invisible wind lines.

Final Note

The equations here aren't substitutes for beauty—they are **tools for resonance.** UED doesn't refute relativity; it **completes it** by including what Einstein missed: **the voice of the field.** With this mathematical backbone, UED transitions from being an intuitive philosophy to a rigorous, testable, and actionable framework for space navigation.

"Motion is not caused—it is allowed by alignment."

— *Sandeep Chavan*

Appendix D: Mathematical Addendum

Ripple Equations, ATT Derivations, and Experimental Formulas

For those who feel the need to translate ripples into ratios.

While the core of Universal Energy Dynamics (UED) is ontological and conceptual, not strictly mathematical, certain formulations can offer clarity, reproducibility, and testability for those trained in quantitative reasoning.

This appendix provides a technical scaffold—a launching pad for experimental research and mathematical development rooted in field logic, ripple behavior, and alignment dynamics.

I. The Ripple Equation (Foundational Form)

UED proposes that ripple events are field disturbances propagating through Structured Vacuum Energy (SVE). Unlike wave equations in conventional physics, UED does not assume a mediumless vacuum or quantized packets.

Instead, ripple propagation obeys a realignment-driven continuity equation:

$$\frac{d^2R}{dt^2} + \gamma\frac{dR}{dt} + \kappa R = F_{\text{input}}(x,t)$$

Where:

- R(x,t): Ripple alignment deviation at point x and time t

- γ: Damping coefficient (field resistance to distortion)
- κ: Realignment stiffness (field's restoring intensity)
- F_{input}: Source function (external or internal disturbance to field)

This equation models how ripples form, dampen, and seek equilibrium—not via quantized excitation, but through dynamic field tension.

II. ATT Derivation — Alignment Threshold Time

ATT (Alignment Threshold Time) is the minimum delay required for a ripple to become observable through alignment with the observer's field.

We define ATT as a function of both field clarity and ripple strength:

$$\text{ATT} = \frac{1}{\rho_r \cdot \chi_o}$$

Where:

- ρ_r: Ripple resonance strength (amplitude × coherence)
- χ_o: Observer's field sensitivity (perceptual tuning)

Interpretation:

- Stronger, cleaner ripples are perceived faster.
- Higher observer alignment (clarity of consciousness or detector sensitivity) lowers ATT.

This relationship shifts our understanding of time from fixed intervals to alignment-dependent perception latency.

III. Ripple Tension Gradient (Force Analogy)

In UED, what we perceive as force is actually ripple tension stored in misaligned SVE. Define ripple tension τ(x) across space:

$$\tau(x) = \frac{dR}{dx}$$

And force-like behavior arises from the ripple tension gradient:

$$F(x) = -\nabla\tau(x) = -\frac{d^2R}{dx^2}$$

This parallels Newtonian force only superficially. Here, no mass or inertia is assumed—only field resistance to misalignment.

IV. Light as Instantaneous Ripple Alignment

Rather than particles traveling through space, light is a field resolution event. In UED, light's apparent speed (c) is an ATT limit under ideal clarity conditions.

If a ripple aligns with the observer in minimal time, we propose:

$$\mathrm{ATT}_{\mathrm{light}} \approx \frac{1}{c}$$

Given that light is perceived at ~3×10^8 m/s:

$$\mathrm{ATT}_{\mathrm{light}} \approx 3.33 \text{ nanoseconds/m}$$

This redefines speed of light not as a motion limit, but as the fastest field reconfiguration perceivable across one meter of structured vacuum.

V. Experimental Design Proposal: ATT Meters

To empirically validate ATT, we propose:

A. Dual Ripple Source Interference Setup

- Emit identical ripple events at known, controlled intervals.
- Measure observer's perception latency via signal detection tools or EEG if human.
- Vary field clutter, ripple coherence, and observer conditions to study ATT shift.

B. Field Resonance Probe

- Controlled disturbances are introduced in an SVE chamber.
- Measure delay in field response across nodes.
- Data yields ripple velocity as a function of field tension and damping.

VI. Structured Vacuum Index (SVI)

Quantify the "clarity" or "tension" of the field per region:

$$\mathrm{SVI} = \frac{\tau^2}{\gamma \cdot \kappa}$$

Where:

- Higher SVI = more stressed or disturbed region (e.g., near black holes or social chaos)
- Lower SVI = more coherent field (e.g., vacuum rest state, meditative states)

This could one day allow field-mapping of real environments, mental states, or even planetary ripple conditions.

VII. Chavan's Limit — The Upper Bound of Perception Alignment

UED proposes a biological/structural ceiling for alignment sensitivity—beyond which no system (biological or synthetic) can resolve ripples.

$$\chi_o^{\max} \leq \frac{1}{\delta_{\text{field}}}$$

Where δ_{field} is the minimum distortion length allowable in the field without collapse.

This defines the perception boundary of any intelligence, giving rise to a limit similar to Planck scale—but derived from consequence resolution, not quantum fluctuations.

VIII. Summary Table: Key Variables & Units

Symbol	Quantity	Interpretation	Units
R	Ripple Deviation	Displacement from alignment	arbitrary (field)
γ	Damping Coefficient	Resistance to ripple	1/time
K	Realignment Coefficient	Stiffness of field response	$1/\text{time}^2$
ρ_r	Ripple Strength	Amplitude × Coherence	1/time
χ_o	Observer Sensitivity	Perceptual tuning capacity	1/time
τ	Ripple Tension	Misalignment intensity	field/m
ATT	Alignment Threshold Time	Delay before ripple perceived	time
SVI	Structured Vacuum Index	Field stress indicator	dimensionless

Final Thought

These equations are not meant to replace all of physics. They are meant to reframe what we measure, why we measure it, and how it matters.

UED invites us to move from mathematical domination to mathematical resonance. From forcing models to aligning understanding.

Appendix E: Experimental Roadmap

Testable Hypotheses and Field-Based Experiments Under UED

A guide for those ready to ripple truth into action.

Universal Energy Dynamics (UED) is not merely a metaphysical proposition or philosophical critique of modern physics. It is a testable, falsifiable, and measurable framework—one that invites experimentalists to observe consequences instead of chasing particles.

This appendix outlines specific experiments, hypotheses, and measurements that can validate or challenge the foundations of UED. No billion-dollar colliders needed. Only attention, precision, and a willingness to see what mainstream models ignore.

I. Hypothesis 1: Time Emerges as ATT (Alignment Threshold Time)

Claim: What we call "time" is not a dimension, but a delay in ripple realignment within Structured Vacuum Energy (SVE). ATT varies based on ripple coherence and observer sensitivity.

Experimental Setup:

- Create controlled ripple disturbances (acoustic, electromagnetic, or mechanical) in structured mediums.
- Measure perceptual or sensor detection latency against ripple clarity (coherence and intensity).

- Vary environmental alignment (e.g., field clutter, harmonic background).

Observable Outcome:

- Higher coherence = lower ATT
- Time perception and detection speed will vary with field alignment, not clock time.
- This will establish that perception time is field-relative, not absolute.

II. Hypothesis 2: Gravity = Ripple Tension Gradient

Claim: Gravity is not a force or curvature—but a gradient of ripple tension in the SVE. Bodies fall because field misalignment invites restoration.

Experimental Setup:

- Construct a sensitive field-mapping platform around a mass distribution (e.g., rotating dense object or layered plates).
- Measure changes in field response using high-resolution interferometry, magnetometers, or scalar tension sensors.
- Compare ripple response with and without mass shifts.

Observable Outcome:

- Tension patterns will form around dense objects.
- Gravitational-like effects will correlate with ripple field distortions, not just mass itself.

III. Hypothesis 3: Light is Ripple Alignment, Not Photon Travel

Claim: Light is not a particle or wave traveling through vacuum—it is a field alignment event occurring when the observer's internal ripples sync with the source.

Experimental Setup:

- Construct an observer-synced light perception chamber.
- Use variable alignment lenses (non-standard filters, field-disrupting materials).
- Track how light is perceived when observer's field is temporarily altered (e.g., with binaural audio, biofeedback loops, or mental state shifts).

Observable Outcome:

- Perception latency or clarity of light will vary with observer's internal ripple state.
- Detection will not always correlate with source intensity—but with resonance conditions.

IV. Hypothesis 4: Misalignment Stores Tension (Force Is Emergent)

Claim: What we call "force" is the stored tension of unresolved ripple misalignment. External force is the consequence, not the cause.

Experimental Setup:

- Build a ripple-loaded environment: generate structured interference patterns in acoustic, fluid, or electromagnetic fields.
- Measure system behavior with increasing ripple distortion.

- Track how much resistance (measured via force probes) emerges before any physical displacement.

Observable Outcome:

- Measurable "force" appears even without motion, purely from tension buildup.
- Resistance correlates to ripple field distortion, not Newtonian inertia alone.

V. Hypothesis 5: ATT-Based Time Dilation Explains Relativistic Anomalies

Claim: All relativistic time effects (twin paradox, GPS corrections, etc.) are better explained by variable ATT, not spacetime curvature.

Experimental Setup:

- Reanalyze existing relativistic data (GPS satellite drift, muon decay rates) using ATT as an explanatory framework.
- Simulate field alignment values to compare predicted delays with actual observational data.

Observable Outcome:

- ATT modeling will yield results matching or exceeding GR predictions—without requiring spacetime curvature.
- Introduces variable perception delay as primary cause of time deviation.

VI. Hypothesis 6: Ripple-Induced Consciousness Alignment

Claim: Conscious awareness increases with ripple coherence. Ripple alignment therapies or exposure to structured SVE fields will show measurable mental clarity and emotional stability.

Experimental Setup:

- Subjects undergo controlled exposure to coherent field conditions (sound frequencies, geometric resonance, guided breath).
- Use EEG, HRV, and cognitive performance tasks to track change.
- Compare with exposure to incoherent, dissonant fields.

Observable Outcome:

- Enhanced ripple coherence correlates with lower ATT, increased focus, and emotional equilibrium.
- Suggests consciousness is field-sensitive, not brain-confined.

VII. Proposed Devices for UED Validation

Device Name	Purpose	UED Functionality
ATT Meter	Detect real-time ripple alignment latency	Measures ATT shifts during field events
SVE Tension Probe	Map ripple field resistance	Reveals invisible field pressure/tension
Field Memory Scanner	Record residual ripple echo after interaction	Tracks field history; analogous to "karma log"
Resonance Amplifier	Enhance field alignment for perception	Aid in perception of subtle ripple events
Ripple Realignment Timer	Measure time to equilibrium post disturbance	Measures field's return to balance

VIII. Guiding Principles for Ripple-Based Science

1. Measure consequences, not entities.
2. Look for coherence before causality.
3. Don't isolate systems—observe ripple feedback.
4. Time is not a constant—it's a local consequence.
5. No result is ever inert. Every test ripples.

Final Note to Experimentalists

You don't need to wait for institutions to validate this. UED is open-source physics for field-aware minds.

You can begin in a lab, a classroom, or even a quiet room with tuned attention. Because ultimately, what validates UED is not a peer-reviewed paper—

It's the moment you perceive the ripple that was always there, and realize: it's waiting to align with you.

Appendix F: For the Silent Reader

A Closing Whisper for Those Who Found Peace, Not Just Ideas.

This page is not for the critic. Not for the scientist. Not for the philosopher. Not even for the seeker.

This is for you— The one who felt something... shift. Not in your thoughts, but in your stillness.

You didn't underline every page. You didn't need to debate. You weren't here to "get" the theory—You were here to feel if it was already true.

And you did.

Perhaps you didn't know what ripple tension was. But you've felt it— In the tightness before a decision. In the echo after a loss. In the pause between breath and thought.

Perhaps you've never spoken of ATT. But you've lived it— In the delay between insight and action. In the moment you just knew... and didn't know how.

Perhaps you've never studied physics. But you've watched a falling leaf and understood gravity better than a textbook could teach. You've felt silence louder than equations. You've trusted alignment over analysis.

And that is enough.

You are not "behind" for not questioning every model. You are not "less" for not challenging Einstein. You are not "wrong" for quietly nodding your way through.

Because in a ripple-based universe, Stillness is understanding.

And the deepest intelligence is not what you say— but what you no longer need to.

So if you read this book without a single argument, without a single urge to annotate, without a single desire to prove—

But instead sat with it, breathed through it, and let it soften something unnamed in you—

Then you are the most advanced ripple this field has met.

Because alignment is not noise. It is resonance.

And you, Silent Reader, have been rippling from page one.

One Last Thought

You won't need to recommend this book. You won't need to defend it. You won't need to teach it.

Just carry its stillness. Its awareness. Its echo.

And in a moment yet to come, someone—perhaps a friend, a child, or a stranger— will ask you something the world cannot answer.

And you will smile. Not because you know.

But because you remember. That the field... remembers through you.

Further Reading & References

Published UED Reports, Related Books, and Einstein's Original Works for Contrast

For those who wish to dive deeper—into the ripples, the questions, and the contrasts—this curated list offers a foundation to continue exploring both Universal Energy Dynamics (UED) and the legacy it responds to.

UED Research & Reports by Sandeep Chavan

These published papers form the scientific and philosophical core of the UED framework:

1. **The Ripple Equation:** Realigning Physics Without Geometry – Introduction to ripple dynamics and tension-based causality.
2. **Time Is ATT:** Rethinking Duration Through Alignment Delay – A revolutionary explanation of time as perception delay in ripple realignment.
3. **The Structured Vacuum:** From Nothingness to Everything – Detailed exploration of SVE (Structured Vacuum Energy) as the fundamental medium of reality.
4. **Chavan's Cosmic Theory of Relativity (CTR)** – A structured-field alternative to Einstein's General Relativity, grounded in ripple tension.
5. **Quantum Ripple Mechanics (QRM)** – Reframing quantum events as ripple interferences, not particle probabilities.

6. **The Law of Consequences (LOC):** A Natural Replacement for Force – Explains how all observable phenomena emerge from ripple consequence logic, not external forces.

Available on Zenodo and other open-access research archives. Keywords: *UED, ripple alignment, ATT, SVE, Chavan Limit*

Books by the Author (UED-Aligned)

These titles elaborate on the implications of ripple-based reality across domains—from physics to philosophy to psychology:

1. **Cracks in the Known** – Book 1 in *The War of Thoughts* series, challenging mainstream physics from first principles.
2. **The Human Game** – Understanding emotion, behavior, and mental health through ripple field logic.
3. **Manifestation:** From Nothing to Everything – A metaphysical journey tracing the first ripple of existence.
4. **The Guiding Light** – A story of mentorship and leadership in alignment with natural consequence.
5. **Live Fearless with Fear** – Reinterpreting fear as ripple dissonance and survival response in the field.
6. **Intelligence on Hotline** – A dialogue with the Source: decoding perception, awareness, and divine intelligence.
7. **Self Help: New Syllabus** – Rejecting habits and productivity hacks in favor of ripple mastery and field awareness.

All titles are available through global distributors and online bookstores.

Einstein's Original Works (For Contrast & Context)

For those interested in comparing legacy frameworks with UED's reinterpretations:

1. **Relativity: The Special and General Theory** – Albert Einstein – Einstein's original public explanation of his theories, foundational yet accessible.
2. **The Meaning of Relativity** – Albert Einstein – A deeper dive into the implications and limitations of spacetime theory.
3. **Ideas and Opinions** – Albert Einstein – Essays reflecting Einstein's philosophical and scientific worldview.

Use these not to worship Einstein—but to honor the questions he made inevitable, and the limitations he openly acknowledged.

Suggested Future Reading

- David Bohm – Wholeness and the Implicate Order
- Niels Bohr – Atomic Physics and Human Knowledge
- Fritjof Capra – The Tao of Physics
- Ilya Prigogine – Order Out of Chaos
- Jiddu Krishnamurti – The Urgency of Change
- Spinoza, Heraclitus, Nagarjuna — For ancient insights that resonate with field philosophy.

Final Note

You don't need to read everything. You just need to read with alignment.

When a book ripples something within—pause. That's the real reading. The rest is just paper.

Acknowledgements

This book was never meant to challenge Einstein. It was meant to complete the conversation he started— The one science left halfway. The one philosophy whispered but never proved. The one the universe has always been having with itself.

To Albert Einstein, with deep respect:

Your curiosity cracked open the sky. Your equations built the bridge. This work simply crosses it, barefoot—with silence instead of chalk.

To my family, who quietly stood in the background, watching me type thoughts the world might not understand— and loving me anyway.

To my students—past, present, and future— You didn't just learn from me. You taught me how to listen again. You showed me where physics ends and life begins.

To the scientists and sages, from Newton to Krishna, Schrödinger to the Buddha— Thank you for making the ripples visible, even when your words were wrapped in paradox or prayer.

To the doubters and disbelievers— You were my mirror. You kept my alignment sharp. And somewhere in your resistance, you preserved my rhythm.

To those who read this book in silence, without needing approval, footnotes, or validation— You are the proof that resonance exists.

To Intelligence itself— not the academic kind, but the cosmic one— subtle, awake, and always whispering from behind the curtain.

You wrote this book before I ever touched the keys. I simply... remembered.

About the Author

Sandeep Chavan is an independent physicist, philosopher, and educator who prefers questions over credentials and silence over applause. With no formal lab, no institutional funding, and no allegiance to academic dogma, he built a new science—not from equations alone, but from consequence, alignment, and the whispering logic of nature.

He is the founder of the Universal Energy Dynamics (UED) framework—a ripple-based understanding of the universe that replaces forces with consequences, particles with field harmonies, and spacetime with structured intelligence.

A lifelong teacher by heart and a researcher by instinct, Sandeep spends most of his time mentoring students, decoding paradoxes, and translating silence into science. His work spans physics, consciousness, human behavior, and cosmology—all viewed through the ripple lens.

He believes truth doesn't need defending—it needs witnessing.

This book is not a conclusion. It's a beginning. Not a theory—but a return to alignment.
You can find his published works, papers, and books globally—but the real alignment starts wherever you begin to question reality... gently.
Visit Sandeep's ORCID Profile:
https://orcid.org/0009-0005-4190-7164

Don't miss out!

Visit the website below and you can sign up to receive emails whenever Sandeep Chavan publishes a new book. There's no charge and no obligation.

https://books2read.com/r/B-A-EPGPC-MHNLG

BOOKS 2 READ

Connecting independent readers to independent writers.

Did you love *Chavan vs Einstein: The Universe Was Never Curved*? Then you should read *TIME – The New Definition*[1] by Sandeep Chavan!

[2]

What if everything you believe about time is backward?

In *TIME: The New Definition*, independent researcher and educator **Sandeep Chavan** invites you to reconsider time not as a ticking clock or a cosmic dimension, but as a **perceptual consequence of misalignment** — a ripple delay born from the interaction between you and the universe.

Drawing from his groundbreaking framework **Universal Energy Dynamics (UED)**, Chavan offers a radical yet intuitive insight:

Time is not something you live in — it's something you generate.

---

1. https://books2read.com/u/mloX9Y
2. https://books2read.com/u/mloX9Y

Through a brilliant fusion of science, consciousness, and first-principle logic, this book dismantles the illusion of time as an objective force. Instead, it introduces **ATT (Alignment Threshold Time)** — the measurable delay that arises when your inner intent meets the outer field out of sync. When you're in harmony, time collapses. When you're in conflict, it stretches. From this vantage point, trauma, boredom, flow, and even deep sleep become navigable terrains of ripple experience.

Across 18 concise yet powerful chapters, *TIME: The New Definition* challenges everything from Einstein's relativity to the Big Bang theory — replacing them with ripple-based logic that restores the **observer to the center of physics**. The book explores:

Why the past and future are just echoes in ripple memoryHow spiritual states like moksha or nirvana align with ATT=0Why sleep, silence, and creativity all dissolve the feeling of timeHow time can never end cosmically, but can be exited spirituallyWhy you are not the timeline — you are the one who generates it

Far more than a scientific proposal, this book is a personal awakening — a field guide for thinkers, seekers, scientists, and skeptics who sense that something about time has always felt... off.

Whether you're a physicist searching for a deeper model, a meditator who's touched timelessness, or simply someone who wonders where all the time goes — this book gives you more than an answer.

It gives you a new clock: **yourself.**

Sandeep Chavan, author of over 100 books and creator of UED, delivers a bold, paradigm-shifting voice for 21st-century science and timeless awareness.

Read more at https://www.sandeepjchavan.com/.

Also by Sandeep Chavan

Chavan's UED research

CMGED Research Report

Chavan's Cosmic Continuum (C^3)

Frequency

Ripples of Reality

Reality Realigned

The Great Realignment

I – The Cosmic Leverage

Chavan's Cosmic Theory of Relativity (CTR)

Chavan's Theory of Light

Feeling Like a Physicist

TIME – The New Definition

Echoes of War: The Russia-Ukraine Conflict and Its Global Ripples

Shattered Lives

The War Economy

Borders Redrawn

Arms and Allies

Collateral Damage

Justice in the Ashes

Somewhere in the Darkness of the Cosmos

The Blueprint of the Universe

The Cosmic Breath

From the Edge of the Universe

The Eternal Sage Chronicles

The Sage and the Sovereign

The Family Pillar

The People's Flame

The Warrior's Soul

The Bridging Mentor

The Guiding Light

The Exodus

The Last Cycle: 2075

The Fundamental Theory of Life & Existence

Energy in Action: The Soul Demystified

The Active Mind

Energy in Connection

Death: The Disconnection

Subconsciousness: The Chosen Disconnection

Desires: The Engine of Life

The Restoration of Human Knowledge

Restoring Bhagavad Gita

The War of Thoughts

Cracks in the Known

The Human Game

The Divine Dilemma

The Original Intent

The Answers

Ultra Modern Life Philosophy

The Duality of Laughter and Tears

Take It Easy

Compare to Accept

Standalone

It's Not AI, It's AHI - Amplified Human Intelligence

The Decision Paradox

The Four Sapiens

Malicious Script of Indian Polity

The IIT Legacy & Global Impact

The IIT Legacy & Global Impact

You are just a Version of Your Original

Win the Game You Didn't Choose

Always Keep Your Bags Packed

Reveal to Shield - Learn the Game of Facades

Engage Beyond Elections
Culture, Identity & Change: The Evolution of Indian Society
The Triangular Dynamics
Never Trust Without Doubt
If You Can't Change the Country, Leave the Country
Live Fearless with Fear
India: The Investment Magnet
Realignment
Powering a Billion Dreams
Universal Skill
Extension: From Brush to Algorithm
Human Side of Einstein
The Future Knows No Lines
Let's Bring the Generations to the Table
Unified Energy Dynamics
A Beat That Never Dies
Strategic Horizons
Right From Scratch
The Forgotten Innovators
Physics Fixing
The Savior Illusion
The Selfish Side
The Age of the Weakest Sapiens
Neural Wars
The Absorption: There Is No Death
Strange Therapist
Bubbles: Become a Drop
Beyond the Blue
Half-Scholar
The Law of Truth
Failure
Philosophy 2.0: A Structural Manifesto
Chaos: A Gentle Guide to the Unresolvable

Self-Help: New Syllabus
Rippling
Game: How to Play the Share Market
Out of Syllabus
Saffronized: The Code Behind the Crown
The Anger
Moksha: The Liberation
Chavan vs Einstein: The Universe Was Never Curved

Watch for more at https://www.sandeepjchavan.com/.